Motor Vehicles, the Environment, and the Human Condition

ENVIRONMENT AND SOCIETY

Series Editor
Douglas Vakoch

As scholars examine the environmental challenges facing humanity, they increasingly recognize that solutions require a focus on the human causes and consequences of these threats, and not merely a focus on the scientific and technical issues. To meet this need, the Environment and Society series explores a broad range of topics in environmental studies from the perspectives of the social sciences and humanities. Books in this series help the reader understand contemporary environmental concerns, while offering concrete steps to address these problems.

Books in this series include both monographs and edited volumes that are grounded in the realities of ecological issues identified by the natural sciences. Our authors and contributors come from disciplines including but not limited to anthropology, architecture, area studies, communication studies, economics, ethics, gender studies, geography, history, law, pedagogy, philosophy, political science, psychology, religious studies, sociology, and theology. To foster a constructive dialogue between these researchers and environmental scientists, the Environment and Society series publishes work that is relevant to those engaged in environmental studies, while also being of interest to scholars from the author's primary discipline.

Titles in the Series

Australian Wetland Cultures: Swamp Thinking, edited by John Charles Ryan and Li Chen

Motor Vehicles, the Environment, and the Human Condition: Driving to Extinction, by Hans A. Baer

Motor Vehicles, the Environment, and the Human Condition

Driving to Extinction

Hans A. Baer

LEXINGTON BOOKS
Lanham • Boulder • New York • London

Published by Lexington Books
An imprint of The Rowman & Littlefield Publishing Group, Inc.
4501 Forbes Boulevard, Suite 200, Lanham, Maryland 20706
www.rowman.com

6 Tinworth Street, London SE11 5AL, United Kingdom

Copyright © 2019 The Rowman & Littlefield Publishing Group, Inc.

All rights reserved. No part of this book may be reproduced in any form or by any electronic or mechanical means, including information storage and retrieval systems, without written permission from the publisher, except by a reviewer who may quote passages in a review.

British Library Cataloguing in Publication Information Available

Library of Congress Cataloging-in-Publication Data Available

ISBN 978-1-7936-0488-0 (cloth : alk. paper)
ISBN 978-1-7936-0490-3 (pbk : alk. paper)
ISBN 978-1-7936-0489-7 (electronic)

Contents

Introduction		1
1	The Political Economy of Motor Vehicles	13
2	Cultural Tropes Associated with Motor Vehicles	45
3	The Political Ecological Impact of Motor Vehicles on Settlement Patterns	65
4	The Impact of Motor Vehicles on the Natural Environment	85
5	The Political Ecological Impact of Motor Vehicles on Health	113
6	Case Studies of Motor Vehicles in Various Places	135
7	Beyond the Automobile	163
8	Creating a Sustainable Transportation System within the Context of an Alternative World System	191
References		219
Index		247
About the Author		251

Introduction

Automobility is a relatively late form of mobility in the human chronicle. In terms getting from one place to another, walking became possible for relatively long distances with the emergence of bipedalism among our early hominine or human ancestors possibly as far back as five or six million years ago. Still later, particularly with the advent of the state, came new forms of mobility facilitated by horses, either ones that were ridden or pulled chariots, wagons, or buggies, and boats of various sorts, particularly canoes, sailboats, and ships. Aside of walking, these other mobilities as well as more recent ones, such as trains, bicycles, automobiles, buses, planes, air ships and balloons, and even crutches and wheelchairs, have relied upon technological devices (Urry 2007). Modern societies have become increasingly mobile societies in which ordinary people in particularly developed societies but also developing societies "are increasingly on the move *and* communicating more to connect with absent others" (Larsen, Axhausen, and Urry 2006:261). For example, whereas in 1950, the PKT (passenger kilometres traveled per capita GDP [in purchasing power parity] world-wide was 3,560 billion, by 2005 it had increased to 38,000 billion. At the extreme end, in North America the PKT for North America alone increased from 1,860 billion in 1950 to 9,120 billion in 2005 (Schaefer, Heywood, Jacoby, and Waitz 2009: 37).

Most mobile technologies that have been developed since the Industrial Revolution rely upon fossil fuels, namely coal, petroleum, and natural gas. Mobile technologies dependent upon fossil fuels, primarily oil today, produce air pollution and carbon dioxide and, in the case of airplanes, carbon dioxide and other greenhouse emissions which are contributing to a global ecological crisis, including global warming and other climatic changes. As Urry (2007:12) observes, mobilities "entail risks, accidents, diseases, trafficking, terrorism, surveillance, and especially global environmental damage."

In this book, while I touch upon the adverse ecological and climatic consequences of trains, buses, trucks, airplanes, ships, and boats, I focus upon private motor vehicles, particularly automobiles, as the central form of mobility for most people in developed societies and an increasing number in developing societies. While private motor vehicles have served largely the more affluent but also sometimes the less affluent sectors of humanity with a relatively rapid and often convenient form of mobility, their negative impacts on the environment, the climate system, settlement patterns, sociability, and health have come to reach a crisis stage in the early twenty-first century, thus warranting moving beyond them as much as possible. Whereas on the one hand automobility provides a flexible form of mobility that expresses both individual identity and group identities, on the other hand it constitutes a "coercive practice that consumes massive amounts of space, requires lengthy commutes, increases dependence among youth, elderly, and others unable to drive or without access to a vehicle, relies upon extensive state surveillance, harms or eliminates environmental options for future generations, and structures patterns of living, working, and playing that often preclude many from more than nominal use of alternatives even where these are available" (Myer 2015:122).

This book seeks to explore the political ecology of motor vehicles in an era of increasing social disparities and ecological crises, the latter which are most manifested in the form of anthropogenic climate change to which motor vehicles constitute a major contributor. While much has been written over the past several decades on the production and consumption of motor vehicles, cultural tropes associated with them, their impact on settlement patterns, the environment, physical and mental health, and social life, most of this literature is not framed in a political ecological perspective. An exception to this observation is Peter Freund's and George Martin's *The Ecology of the Automobile* (1993). However, much has happened since the publication of their now classic book, including a better awareness of the impact of motor vehicles on climate change and the rapid expansion of the culture of automobility to developing countries, particularly China, India, Brazil, and South Africa as well as former Soviet-bloc countries.

As a critical anthropologist, I treat political economy and political ecology as inseparable. As Howard L. Parsons (1977: xii) argues in *Marx and Engels on Ecology*, "Economy is a matter of ecology: it has to do with production and distribution of goods and services in the context of human society and nature. . . . [It recognizes that] under the ecological practices of monopoly capitalism, the natural environment is being destroyed along the social environment." My book seeks to promote the political ecology of mobility or transportation studies, with a focus on private motor vehicles. A political ecological perspective recognizes that motor vehicles, with their internal combustion engine reliant upon oil, perhaps more than any other machine embody the social structural, cultural, and environmental contradictions of the capitalist world system. Motor vehicles have had major impact upon patterns of consumption, settlement (e.g., suburban sprawl), traffic congestion, mass transportation, social relations, public policy, the environment, and health. Motor vehicles touch the natural world at four points: (1) the extraction of raw materials for their production which results in the depletion of natural resources; (2) their manufacture which contributes to environmental pollution and greenhouse gas emissions; (3) consumption which further contributes to environmental pollution and greenhouse gas emissions; and (4) their disposal once they have outlived their usefulness. While automobility has provided capitalist enterprises with massive profits and access to markets, it has taken a tremendous toll on the ecosystem and the climate, settlement patterns, social life, and both physical and mental health.

I have come to the study of private motor vehicles as both a critical anthropologist and historical social scientist and a scholar activist committed to social justice and environmental sustainability. At a personal level, as I have lived in different countries and traveled to others, I am continually reminded of the extent to which particularly automobiles govern people's movements and lives in one way or other. I have become increasingly annoyed that I must wait for automobiles to pass when trying to cross streets and that I have to look out for them when cycling. I am also struck by the few pedestrians I see on streets in contrast to passing motorists in their automobiles or the massive number of parked automobiles that I see while walking or cycling from one place to another. Generally, on a quiet street, one sees many more automobiles than people. A fictional visitor from another world might conclude that the primary inhabitants of large swathes of the Earth are automobiles and other motor vehicles. Along with industrial pollution, private motor vehicles have transformed many cities around the world into environmental disaster areas accompanied by congestion, destruction of land, a wide array of health problems, and social disruption. Along with their environmental impacts, motor vehicles have had major impacts on not only patterns of consumption but also energy utilization, settlement patterns, social relations, public policy, and congestion, all of which in one way or another affect health. Orssatto and Clegg (1999:264) boldly assert the automobile industry constitutes the "economic sector most emblematic of modern times and of the polluting consequences of modernity." In a somewhat related manner, Sperling and Gordon (2010:1) maintain that automobiles constitute "one of the greatest man-made threats to human society."

Jeremey Packer (2008:12) asserts that "[c]ivil societies of the west are societies of automobility." While this is certainly true of particularly the United States, Europe, and the settler societies of Canada, Australia, and New Zealand, in today's world of increasing corporate globalization it increasingly holds true for much of the developing world, including the parts which never fell directly under the hegemony of European colonialism. Ward's research reveals that based upon government-reported registrations and historical vehicle-population trends, there were 1.015 billion registered motor vehicles in the world in 2010 (Sousanis 2011). Voelcker (2014) maintains that there were 1.2 billion motor vehicles on the world's roads in 2014. He maintains that the Ward's calculation includes passenger cars, light-, medium-, and heavy-duty trucks, and buses, but does not include off-road vehicles, such as farm vehicles, and heavy construction and all-terrain vehicles.

Whatever the exact number of motor vehicles in the world today, their number can be expected to continue to rise as certain developing countries, particularly China, India, Brazil, Russia, and others imitate developed societies in their adoption of a culture of automobility. Sperling and Gordon (2010:4) project that there will be over 2 billion motor vehicles, at least half of them cars, by 2020 while the International Monetary Fund projects the presence of some 3 billion cars in the world by 2050 given the current pace of car production (Richter 2010:118). The latter projection may be unlikely given the impact of climate change over the course of next several decades which is likely to have grave effects on the global economy and life in the current era. Part and parcel of these impacts, motor vehicles have had and will continue to affect patterns of consumption, settlement, traffic congestion, mass transportation, social relations, the environment, and health. They constitute a major contributor to not only air and noise pollution but also to greenhouse gas emissions and ultimately to climate change as well as fatalities, injuries, stress, social isolation, and obesity.

The private automobile constitutes the ultimate commodity for many people around the world, indeed a *killer commodity* and a *death machine*, around which numerous other industries operate, including the oil, accessory parts, highway construction, insurance, and even health care industries. As Matthew Paterson (2007: 92) astutely observes, "[a]cross a wide range of political/economic discourses the car has been seen to play a fundamental role in the promotion of economic growth in the twentieth century [as well as the twentieth-first century], and thus in the reproduction of capitalism as a system." The state has served as a staunch ally of automobile companies by facilitating road construction, allowing the systematic neglect and downplaying of public transportation, fiscal policies which essentially privilege the car over other forms of transportation, and rescue the car industry from economic collapse (Paterson 2007:115).

Various approaches to automobility have appeared over the course of the past 15–20 years. At some level, they recognize that automobiles became a hegemonic force during the 20th century and continue to occupy this role in the present century. As Paul Gilroy observes,

> The twentieth century was the century of the automobile, of automobility and mass motorization. Commerce in motor vehicles still constitutes the overheated core of unchecked and unsustainable consumer capitalism, but the impact of car culture extends far beyond those buoyant commercial processes.... Novel and damaging patterns created by motorization have profoundly altered the political economy of everyday life. (Gilroy 2001:81–83)

In a somewhat different vein, Steffen Boehm et al. maintain that:

> Automobility is one of the principal socio-technical institutions through which modernity is organized. It is a set of political institutions and practices that seek to organize, accelerate and shape the spatial movements and impacts of automobiles, while simultaneously regulating their many consequences. It is also an ideological ... or discursive formation, embodying ideals of freedom, privacy, movement, progress and autonomy, motifs through which automobility is represented in popular and academic discourses alike, and through which its principal technical artefacts—roads, cars, etc.—are legitimized. Finally, it entails a phenomenology, a set of ways of experiencing the world which serve both to legitimize its dominance and radically unsettle taken-for-granted boundaries separating human from machine, nature from artifice and so on. Together these apparently diverse strands comprise an understanding of automobility that is irreducible to *the* automobile. (Boehm et al. 2006:3)

Boehm et al. (2006:7) adopt a Foucaultian take on automobility which views it is part and parcel of regimes of power which has "gone hand in hand with a deepening of state power." They recognize that the regime of automobilty is high dependent upon oil, access to which may impel governments to go to war.

Sociologist John Urry (2008:344) adopts a conception of automobility which is more consistent with the approach that I emphasize in this book, an eco-Marxian or eco-socialist one. He delineates six key aspects of cars and the associated culture of *automobility*:

- "the quintessential *manufactured object* produced by the leading industrial sectors and the iconic firms within 20th-century capitalism" and one might add early 21st-century capitalism;

- "the major item of *individual consumption* after housing which provides its owner/user through its sign-values (such as speed, security, safety, sexual desire, career success, freedom, family, masculinity)";
- "an extraordinarily powerful *complex* constituted through technical and social interlinkages with other industries";
- "the predominant global form of 'quasi-private' *mobility* that subordinates other mobilities of walking, cycling, travelling by rail and so on" and which defines work and recreation;
- "the dominant *culture* that sustains major discourses of what constitutes the good life"; and
- "the single most important cause of *environmental resource-use*" which serves as a major source of pollution, carbon dioxide emissions, health problems, as well as an indirect contributor to warfare, particularly in efforts to ensure oil supplies (Urry 2004:25–26)

I examine private motor vehicles, particularly automobiles, within the larger context of the capitalist treadmill of production and consumption. Motor vehicles, with their internal combustion engines, perhaps more than any other machine embody the social structural, cultural, and environmental contradictions of the capitalist world system. What Foster (2002:98) terms "automobilization-industrialization" includes not only automobile and other motor vehicles but also the "glass, rubber, and steel industries, the petroleum industry, the users of highways for profit (such as the trucking firms), the makers of highways, and the real estate interests tied to the urban-suburban structure." This elaborate complex is a source of capital accumulation and economic growth as well as political and cultural power.

Motor vehicles have major impacts upon consumption, settlement patterns (e.g., urban sprawl), traffic congestion, mass transportation, social relations, public policy, the environment, the climate system, and health. James Flink (1988: viii) contends that the rise of the automobile industry and a massive network of roads are "central to the history of the advanced capitalist countries in the twentieth century." Motor vehicle production and consumption have been a central component of developing capitalist economies as well as post-revolutionary or socialist-oriented ones and this process continues and expands in the present century. The production and consumption of automobiles became a major component of Fordism, a term coined by Antonio Gramsci to designate the twenieth-century vision of mechanized production coupled with the mass consumption of standardized products.

John M. Myer refers to automobility as a material practice that he breaks down into two broad constellations. The first component encompasses the "roads and highways, parking structures, driveways and garages, traffic laws and enforcement, gas stations, refineries, dealerships and manufacturers, transformed urban, suburban, and rural forms and landscapes and many other material components that are integral to driving an individual automobile" (Myer 2015:121). The second component entails "imaginary and attitudes toward driving and car culture, perceptions of space and speed and of the relation between technological innovation and cultural change, and the ways these interact with social relations of gender, race, and class, as well as political discourse" (Myer 2015:121). In contrast to Urry's conception of automobility, a weakness of Myer's conception is that while it recognizes the political aspects of automobility, it overlooks the political economy of automobility, including how embedded automobiles are or have been in both capitalist and post-revolutionary societies, such as the Soviet Union, the various Eastern European countries of the former Soviet-bloc, and China in the current era.

In many ways, the culture of automobility is deeply embedded within the parameters of the capitalist world system and serves to reinforce it, despite its numerous contradictions both at the socioeconomic and ecological levels. As Freudenda-Pedersen observes,

> The individual knows that automobility pollutes, makes noise and kills, but this does not mean that transportation habits are altered. The late modern everyday life is time-pressured and filled with many different types of risks (greenhouse effect, traffic jams, obesity, etc.) about which we are constantly compelled to make our minds up. Living with these risks have become a normal condition and state of life in which the individual has learned to navigate within and around with the use and implementation of different tools. (Freudendal-Pedersen 2009:34)

In this book, I seek to highlight many of the environmental, social, and health consequences of humanity's increasing reliance on motor vehicles, particularly private automobiles. As Stilwell (1992:35) so aptly argues, the "private automobile serves the profit interests of the automobile, oil, tyre and ancillary industries, but it is far from being the best form of urban transport in terms of cost, health and ecological sustainability." Bearing this thought in mind, I argue that we need as a species to move beyond motor vehicles as much as possible but that such an effort will have be part and parcel of creating an alternative world system based on social justice, democratic processes, and environmental sustainability.

Chapter 1 (The Political Economy of Motor Vehicles) examines the political economy of motor vehicles, highlighting the social relations of production, access to and control over resources required for their manufacture and operation, and the power relations between motor vehicle companies, the highway lobby, and the state in various social formations or societies. The production and consumption of motor vehicles, particularly automobiles, became a major component of Fordism which propelled the automobile from a luxury commodity into a mass commodity. Historically, motor vehicle production has been concentrated in Europe, North America, and Japan with production and consumption on the rise in developing countries, particularly South Korea and more recently China, India, Brazil, and South Africa. Historically, there has existed a close nexus between the motor vehicle industry and the state, one that has played itself out differently historically and cross-nationally. In the United States, a powerful highway lobby consisting of motor vehicle, oil, and tire companies, highway construction companies, and motorist clubs, have heavily influenced party politicians to pass legislation the promotes the culture of automobility. Nazi Germany constituted another prime example of the close ties between the motor vehicle industry and the state. With the overproduction of motor vehicles in developed countries, motor vehicle companies have sought new markets in the Third World and, with the collapse of the Soviet bloc, Eastern Europe. The Chinese car industry consists of state companies as well as various joint operations with foreign companies.

Drawing upon the previous chapter which recognizes that motor vehicle companies, along with a powerful highway lobby and the state, in both capitalist and post-revolutionary societies serve as the drivers of the culture of automobility, chapter 2 (Cultural Tropes Associated with Motor Vehicles) examines several cultural tropes associated with motor vehicles around the world. These include the automobile as a symbol of social status and self-worth; the automobile as a symbol of personal freedom, convenience, excitement, and adventure; the automobile as a compensation for alienation in the workplace and social life; the automobile as a symbol of gender and sexuality; the automobile as a symbol of national and ethnic identity; the automobile as a living room and work space; and the glorification of private motor vehicles in novels, films, and songs.

Chapter 3 (The Political Ecological Impact of Motor Vehicles on Settlement Patterns) examines how motor vehicles have contributed to urban and suburban sprawl, congestion, and social life around the world. The rise of industrial capitalism contributed to the rapid growth of cities around the world, initially in developed countries but more recently in developing countries. While suburban sprawl initially grew in tandem with suburban trolley and rail lines, it exploded particularly with the advent of the mass production of automobiles beginning in the early twenieth century. The motor vehicle industry and highway lobby have served as formidable forces in the deterioration of public transit systems and impeding their revival in numerous cities. Lewis Mumford (1964) argued that federally-funded highway programs in the US contributed to the creation of a "one-dimensional transportation system." Car dependency is particularly pronounced in North American and Australian cities but is quickly spreading to many cities in developing countries. Motor vehicles have also contributed to tremendous congestion problems in which they make their way across large cities such as Los Angeles, Houston, Mexico City, Sao Paulo, Beijing, Shanghai, Jakarta, and Bangkok in a long series of starts and stops.

Chapter 4 (The Impact of Motor Vehicles on the Environment) examines the contribution of motor vehicles to the depletion of natural resources, environmental degradation, air pollution, greenhouse gas emissions, and thus climate change. Along with industrial pollution, motor vehicles have contributed to air pollution particularly in urban areas around the world, particularly more recently in developing countries. Motor vehicle emissions include sulphur oxides, nitrogen oxides, carbon monoxide, hydrocarbon, volatile organic compounds, toxic metals, lead particles, particulate matter, and carbon dioxide, a greenhouse gas which is a major course of anthropogenic climate change. Motor vehicles also constitute a major source of noise pollution which reportedly contributes to both physical and mental health problems. As a point of comparison, I also briefly examine the impact of other modes of transportation, namely trains, airplanes, and ships on the environment.

In chapter 5 (The Political Ecological Impact of Motor Vehicles on Health), I discuss the political economic production of fatalities and sickness resulting from the manufacture and utilization of motor vehicles. When automobiles first began to appear, various public health reformers viewed them as a panacea to the manure and urine deposited by horses in cities. In contrast, there is now much evidence that motor vehicles are a major source of various health problems, including deaths and injuries. Heavy reliance on automobiles as a mode of transport contributes to less exercise in the form of walking and cycling and thus to obesity and various cardiovascular diseases. Motor vehicle pollution contributes to respiratory and cardiovascular disease as well as cancer. I also examine the impact of motor vehicles, particularly automobiles, on mental health, particularly in contributing to social isolation, road rage, and drunk driving. Automobile transportation discourages social interaction which at least sometimes occurs in the process of taking public transit and leads to social isolation in that most motorists, especially in developed societies, drive alone. Various studies have demonstrated that automobile drivers, particularly in conditions where traffic congestion is moderate to light, may exhibit "road rage" because they feel frustrated by their inability to reach their destinations quickly. Generations of children in particularly developed societies have internalized car dependence as their parents chauffeur them to and from school and to various activities. Elderly people with impaired vision and reflexes often continue to drive because the automobile is necessary to maintain social connections with friends and family scattered about urban as well as rural areas.

Chapter 6 (Case Studies of Automobiles in Various Places) highlights that from an anthropological or even sociological perspective, what is needed is a cross-cultural and cross-national examination of motor vehicles, one which Daniel Miller (2001) promoted in his anthology on *Car Cultures*. In developed societies, private automobiles, taxicabs, local and long-distance buses, motorcycles, motor scooters, motor bicycles, a wide assortment of trucks, and recreational vehicles account for the better bulk of motor vehicles. In developing countries, while most of these exist as well, one finds various other types of motor vehicles, such as the auto rickshaw, which are not generally found in developed societies. While an obvious feature of American cars has been their size, automobiles and other motor vehicles in our countries, including European ones and Japan, tend to be smaller, often considerably smaller. In the spirit of contributing to a cross-cultural and cross-national examination of motor vehicles, adopting an explicitly political ecological perspective, this chapter presents four case studies of automobiles in selected places, Germany (both West and East), Australia, Brazil, and Cuba. In the case of Germany, I examine the development of two "people's cars," namely the Volkswagen promoted by Adolf Hitler and the Trabant, a rough counterpart to the Volkswagen, in the German Democratic Republic, the latter of which had serious negative environmental side-effects. I also examine at some length automobility in Australia, where I have lived as a transplanted American for over fourteen years, which manifests various similarities and differences with automobility in the United States. This chapter also includes case studies of the rise of automobility in Brazil and how 1950s cars have been kept alive in Cuba and touches upon the need for more case studies of automobility in places such as Africa, China, and India.

Chapter 7 (Beyond the Automobile) examines various critiques of private motor vehicles and systemic-challenging transitional reforms that could overcome the present and growing auto dependence characteristic of the capitalist world system. Despite the existence of massive corporate and government support for the on-going use of automobiles, various counterhegemonic efforts to resist the automobilization of society that emphasize the importance of public transport, cycling, and walking. Environmentalists and other social activists began to challenge the pollution, health hazards, traffic congestion, urban and suburban sprawl, destruction of neighborhoods, and fragmentation of social life resulting from motor vehicles and highways in the 1960s and 1970s, a period of social ferment on many fronts around the world. An increasing number of scholars over the course of the past two decades or so have made suggestions for reducing reliance on automobiles or at least mitigating their worst impacts. A new urbanism that seeks to make cities more liveable and environmentally sustainable has emerged around the world and has begun to penetrate urban planning, calling for better public transport and less dependence on automobiles. Furthermore, there is the issue of connecting small towns and rural areas with cities in other ways than heavy reliance upon automobiles. Reliance upon public transport rather than automobiles contributes to some walking and cycling constitutes even more intense form of exercise.

In chapter 8 (Creating a Sustainable Transportation System within the Context of an Alternative World System), I argue that ultimately moving beyond the automobile can only be adequately addressed through the creation of an alternative world system based upon meeting human social needs, highly democratic processes, environmental sustainability, and a safe climate. In capitalist societies, "time is money," and this dictates rapid movement between places. Conversely, in a more leisurely paced world based on eco-socialist principles, people might find slower train travel to be a time to slow down by reading, chatting with fellow passengers, enjoying the passing countryside, reflecting, and even sleeping. A more sustainable form of vacationing or holidaying would entail trips much closer to home, by train or bus, if possible rather than to distant places by either plane or car. The global economic, ecological, and climate crises that are by the by-products of a capitalist world system requires that we re-examine much of what we do in terms of work and leisure, what we eat and consume in general, in what sort of dwellings we reside, and how we move about our planet. In addition to rejecting, a simpler way would also entail a disposal of or minimization in the use of private motor vehicles and reliance on alternative modes of transportation, including simple walking and cycling.

Chapter One

The Political Economy of Motor Vehicles

Over the course of the twenieth century and continuing into the twenty-first century, automobiles and motor vehicles in general have become part and parcel of what sociologist George Martin terms an "auto formation":

> An auto formation is a platform of the emerging global political economy, which is driven by another round of time and space compression—in the turn-over of commodities, in the reach of the corporation, and in the locations of workers, materials, and markets. Increased automobility, especially in the form of light trucks and vans, is a significant part of this restructuring for it supports one of its principal pillars—outsourcing Simultaneously, the crystallization of an auto social formation is diffusing local landscape conflicts and escalating global environmental problems. (Martin 2002:5)

Indeed, the motor vehicle-oil complex constitutes a major sector of the global economy. In 2016, Volkswagen with an annual revenue of $237 billion was the seventh largest Fortune 500 company and Toyota with an annual review of $226 billion was the eighth largest Fortune 500 company (Fortune 2016). In terms of petro-chemical companies, China National Petroleum ranked 3rd with an annual revenue of $299 billion, Sinopec Group ranked 4th with an annual revenue of $294 billion, Royal Dutch Shell ranked 5th with an annual revenue of $272 billion, Exxon Mobil ranked 6th with an annual revenue of $246 billion, and Beyond Petroleum (BP) ranked 10th with a total annual revenue of $226 billion.

PRODUCTION AND CONSUMPTION

A steam-powered automobile was publicly tested in England in 1801, but steam proved not to be a satisfactory form of power (Wolf 1996:52). Ironically, bicycles, which were first manufactured in France in 1860, paved the way for automobile manufacturing. Much of the technology used in automobile manufacture, such as ball bearings, chain drives, gearing systems, and pneumatic tires, was used in bicycle manufacture (Bardou et al. 1982:6). In 1890 the Peugeot Company in France manufactured a small automobile powered by a one-cylinder German Daimler engine. In the 1890s, German and French engineers pioneered the use of internal combustion engines for powering automobiles. In contrast, Americans turned to internal combustion models after a long period of experimentation with steam-powered and electric motor vehicles. Historically automobile manufacturers have often used racing cars to test not only automobiles but also their component parts, such as engines and tires. In 1895, at the dawn of the automobile era, Benz in Germany and P&K&L and Peugeot in France manufactured almost all of the cars in the world (Kingsley and Urry 2009:31). US production of automobiles began to supersede that of France by 1904–1905. The first three decades of the twenieth century propelled the United States ahead of France and Germany in terms of the manufacture of automobiles. Whereas there were only around 8,000 registered automobiles in 1900 in the United States, this number increased to 485,000 by 1910 and then had skyrocketed to over 23.1 million on the eve of the Great Depression in 1929 (Seiler 2008:36–37). Angus (2016:132) reports: "Sales of cars, trucks, and buses soared from 4,000 in 1900 to 1.9 million in 5.34 million in 1929, by which time automobile manufacturing was the country's largest industry. In addition to selling millions of cars, the industry itself was a huge market for steel, glass, and rubber."

Initially, however, American cars were viewed as a luxury item for rich people that physically endangered ordinary people. Indeed, farmers in New York State and Wisconsin sought to scare motorists away with their shotguns (McCarthy 2007:11). Farmers along with small-town residents, suburbanites, and even urbanites regarded automobiles as dangerous and symbols of arrogance (Kline and Pinch 1996: 768–769). While president of Princeton University, Woodrow Wilson in 1906 reportedly stated that "nothing has spread socialistic feeling in this country more than the use of automobiles. To the countryman they are a picture of arrogance of wealth with all of its independence and carelessness" (quoted in Seiler 2008:37). Similar sentiments emerged in Paris in the late nineteenth and early twentieth centuries where tensions between those who owned automobiles and those who did not. Automobile ownership became "associated with a worldly West Paris made up of well-off owner, including some who made themselves known through the participation in the *Premier Congres international d' automobilisme*, held in 1900 at which nearly 500 motorists came to show off their prize possessions symbolizing their elite status (Flonneau 2006:93).

The earliest production of motor vehicles focused not on ones powered by the internal combustion engine but electric power. Literally thousands of electric automobiles, taxicabs, and delivery trucks traversed US cities in the late nineteenth and early twentieth centuries. The Electric Vehicle Association of America began in 1910 to bring automobile manufacturers and utility companies together to promote battery-operated automobiles (Black 2006:4). None other than Henry Ford worked as an engineer at the Edison Illuminating Company before he went on to form the Detroit Automobile Company in 1899 which collapsed because of a dispute between himself and his investors when he insisted on perfecting his products before mounting a campaign to sell them (Black 2006:99). In creating the Henry Ford Motor Company in 1901, Ford planned first to manufacture perfected racing cars which would serve as the prototypes for mass-produced private automobiles. In forming a partnership in 1914, Edison and Ford pledged that they would provide every US household with a $600 electric car. The demise of the electric car occurred during World War I. Under the tutelage of Henry Ford, automobile production entailed a corporate vision of mechanized production coupled with the mass consumption of standardized products. Fordism came to refer to not only the mass production of motor vehicles but any commodity. In 1926–1927 Ford transferred his manufacturing operations from Highland Park to what became the huge River Rouge complex where foremen came under the surveillance of "servicemen" or musclemen who had been criminals and retired athletes patrolled the factory for all sorts of perceived irregularities, ranging from alleged union organizing to watching for defective machinery and safety violations (Coopley and McKinlay 2010:114–116).

In contrast to the United States, World War I created a lull in the production of automobiles in Europe (Bardou et al. 1982: xiv). Opel introduced assembly-line production of automobiles in Germany in 1924 (Sachs 1992:37). Table 1.1 below depicts production of the ten world's largest automobile companies in 1913, shortly before the outbreak of World War I, all of which were situated in the United States.

Los Angeles County quickly became the leading center of automobile ownership in both the United States and the World. By 1907, the United States had taken the clear lead in the culture of automobility, with 143,200 registered automobiles, in contrast to Britain with 63,500; France with

Table 1.1. Ten Largest Automobile Companies in the World, 1913

Company	Automobiles Produced
Ford (US)	202,667
Wiley-Overland (US)	37,442
Studebaker (US)	31,994
Buick (US)	22,666
Cadillac (US)	17,284
Maxwell (US)	17,000
Hupmobile (US)	12,543
Reo (US)	7,647
Oakland (US)	7,030
Hudson (US)	6,401

Source: Adapted from Bardou et al. (1982:74).

40,000; Germany with 16,214; and Italy with 6,080, one that it still retains to this day (Bardou 1982:20). By the 1920s the Ford Motor Company was manufacturing over a million automobiles a year (McCarthy 2007:55). Indeed, Ford's Rouge factory in Dearborn constituted the "greatest and best-publicized example of vertical integration in twentieth-century industrial America" (McCarthy 2007:58). Some 10 million Model T Fords had been produced but they were replaced by the Model A Ford in 1927.

At the outbreak of the Great Depression in 1929, one in five Americans owned an automobile. According to Ling (1990: 75), "In Detroit, the ratio was even higher: a motor car for every four residents; in Los Angeles, it was even higher: a car for every three inhabitants. No European nation even approached this level of mass car ownership until the mid-1960s."

Motorcycles began to appear in the early twentieth century in Europe and North America. Until the advent of the Model T Ford, they were considerably cheaper than automobiles and could achieve high speeds (Packer 2008:116). Motorcycles came to be used extensively by police departments in the United States during the 1920s and 1930s. As automobiles grew in popularity, motorcycles came to be viewed as a recreational vehicle. In contrast to many motor vehicle manufacturers in the United States, France, and Germany during the first three or four decades of the twentieth century, in a manner characteristic of the shift from competitive to monopoly capitalism, "their number had been reduced to about half-a-dozen mass producers plus a few catering for the prestige luxury market" (Law 1991:6). While there was a lull in the production of automobiles during World War II in various countries in Western Europe, including Britain and Germany, production began to climb afterwards. A second wave of automobilization occurred after the end of World War II. In the case of the United States,

> From 1947 through 1960 the motor vehicle, petroleum, and rubber industries were responsible for one-third of plant and equipment expenditures in manufacturing. Consumer outlays on automobiles and parts, gasoline, and oil rose from 6.5 percent of total expenditures to 9 percent during the same period. By 1963-66 one of every six business enterprises was directly dependent on the manufacture, distribution, servicing and use of motor vehicles; at least 13.5 million people, or 19 percent of total employment, worked in "highway transport industries." (DuBoff 1989: 102)

By 1950, world production of cars totalled 10,017,000. The breakdown was: US, 8,003,000; Britain 783,000; Canada 390,000; France 357,000; Germany 306,000; and Italy 127,000 (Mosey 2000:51). Motor vehicle production followed in classical form the evolution from competitive capitalism to monopoly capitalism. In 1955, the Big Three—Ford, General Motors, and Chrysler—accounted for 95 percent of US automobile sales (McCarthy 2007:101). Quinn reports:

> The remaining four "minors,"—Packard, Studebaker, Nash, and Hudson—did survive in their original form through the 1950s. . . . [Nash and Hudson] merged to form American Motors, a company which survived largely as an American assembler for the French company Renault before it was purchased by Chrysler. (Quinn 1988:65)

This monopoly was undermined beginning in the 1960s when many Americans began to opt for foreign-manufactured automobiles, starting out with the Volkswagen Beetle. US motor vehicle production peaked in 1965 when the industry manufactured 11.1 million automobiles, trucks, and buses (Penna 2015:262).

Production of automobiles also skyrocketed in the defeated countries of World War II, namely Germany and Japan. Particularly after World War II, Japan adopted large-scale motorization, with the number of motor vehicles skyrocketing from some 142,000 in 1945 to 922,000 in 1955 (Garrison and Levinson 2014:298). Japanese motor vehicle companies nearly tripled their automobile production three-fold between 1970 to 1988, increasing production from some three million to eight million per annum (Minchin 2007:328). The oil crisis of 1979 spurred by political upheaval which culminated with the ouster of the Shah of Iran shifted US consumption of automobiles from large ones such as those that historically US motor vehicle companies manufactured to smaller ones that European and Japanese companies had been manufacturing, particularly since the end of World War II. In the early 1980s,

Japanese automobile companies took the lead in production but US automobile companies regained the lead in the mid-1990s, only to subsequently lose the lead. In contrast to traditional Fordist methods which relied upon workers carrying out unskilled and repetitive tasks, Japanese companies introduced "flexible" or "lean" methods in which workers conducted a variety of tasks, looked out for defects, and thus improved the quality and reduced production costs.

The number of automobiles in Japan increased from approximately 142,000 in 1945 to some 922,000 in 1955 (Garrison and Levinson 2014:298). Although it almost collapsed in 1950, Toyota rose like a phoenix and eventually joined General Motors and Volkswagen as a member of the Global Big Three. Although Nagoya is the regional headquarters of the Toyota Motor Corporation, the company's real center is stated about 35 miles away in Toyota city, which may be the "largest company town in the world" today (Keller 1993:56). Elsewhere on Honshu Island, one finds Mazda headquartered in Hiroshima and Honda and Nissan headquartered in Tokyo.

In the wake of the oil crisis of 1973, many Americans shifted from large gas-guzzlers to smaller and more fuel-efficient Japanese and South Korean models. In time some of these grew size-wise and in fuel consumption. For example, when the Honda Accord was released in 1976, it weighed 2,000 lbs, had a 1.6 litre, 68-horsepower engine which got 46 miles per gallon on the open highway (Sperling and Gordon 2010:19). In contrast, the 2008 Accord was 78 percent heavier and had an engine that was nearly four times as powerful, meaning that it got only 29 miles per gallon on the open highway. At the dawn of the twenty-first century, the SUV and the Hummer as "two giant steps backward for fuel efficiency, passenger safety, and inconspicuous consumption" became the "undisputed Kings of the Road in America" but also came to appear in other countries around the world (Dery 2006:223).

Like Americans, Swedes have shifted toward purchasing automobiles with more horsepower. According to Hagman (2010:26), the "six top selling models in 1968 (Volvo 144, Saab V4, VW 1300, Opel Rekord, Volvo Amazon and Opel Kadett), which together counted for almost 60 percent of the total market for new cars, had a weighted average of less than 70 hp, while the eight top selling models in the first half of 2008 (Volvo V/C70, Saab 9-3, Volvo V50, VW Golf, Saab 9-5, Audi 4, Ford Focus and VW Passat), which together had about one third of the total market, had a weighted average of more than 140 hp."

In 2014 China manufactured 19,919,795 automobiles and 3,803,095 commercial vehicles, making it by far the world's leading motor vehicle manufacturer, outstripping both Japan and the United States (International Organization of Motor Vehicle Manufacturers [OICA] 2014). However, China's ability to manufacture world-class motor vehicles relies on technology transferred from the United States, Europe, and Japan. These statistics must be used with some caution. For example, the number of commercial vehicles manufactured in the US in 2014 supposedly greatly exceeded the number of cars manufactured. However, sports utility vehicles are included in the figure for commercial vehicles and many Americans drive light-weight trucks for regular driving purposes rather than commercial purposes. In the United States as well as other developed countries, including Australia, motorists have shifted from cars per se to light trucks, including SUVs, vans, and pickup trucks. However, it is important to note that these figures include only cars and light trucks, but not other types of passenger motor vehicles, including motorcycles, mopeds, motorized bicycles, recreation vehicles or camper vans, and racing cars. As Martin (2002:6) observes, an auto social formation is also "accompanied by an explosion in the use of all form and manner of individualized, motorized off-road transport—snowmobiles, dune buggies, All-Terrain Vehicles, mountain motorcycles, and swamp buggies."

While this book focuses largely upon private motor vehicles, commercial vehicles, which include trucks, along with other transportation modes, which transport goods either long distances, moderate distances, or short distances, as well as buses must be considered in considering the impact of motor vehicles upon settlement patterns, social life, the environment, and health. A United Nations report states:

> Goods transport accounts for 10 to 15 percent of vehicle equivalent kilometres travelled in urban areas and have been linked to the externalities of congestion and air and noise pollution. Evidence indicates that a high-income city in Europe generates about 300 to 400 truck trips per 1000 people per day and 30 to 50 tons of goods per person. Freight movement is largely driven by diesel powered cargo vessels, trucks, and trains and while diesel engines are more energy efficient as compared with petrol, they contribute significantly to GHGs and other short-lived climate pollutants particularly black carbon, impacting therefore also on public health. Despite the significance of goods transport in the urban environment, it has received relatively little attention from policy makers and planners. (United Nations 2015:3)

Until recently, the United States had been the largest producer of motor vehicles in the world, exemplified by the fact that in 2004 alone it manufactured 11.96 million of them (Rutledge 2006: xi). It is also the most highly motorized nation in the world, with 834 registered vehicles per 1,000 people, over 50 percent higher than Western Europe (Rutledge 2006:13). Nevertheless, as Kovel observes,

> Looming overcapacity hangs over automobile industries, as it does for capitalist production in general, with the ability to make over 80 million cars a year, and but 55 million or so able to be sold. Those unrealized 25 million vehicles are a giant splinter in the soul of capitalism, and the goad to endless promotion of automobilious values. (Kovel 2007:76)

Production and Consumption of Motor Vehicles in the USSR and Eastern Bloc Countries

The first Five Year Plan of 1928–1933 under Stalin included the creation of a motor vehicle industry, but it was only under the second and third Five Years Plans that mass production made a sizeable increase (Nieuwenhuis 2014:47). While much less so than in Western societies and Japan, motor vehicle production underwent a significant increase in the Soviet Union following World War II. According to Nieuwenhuis (2014:47), over time a "carefully designed range of centrally planned cars was developed and built in the Soviet Union, starting with small cars such as the ZAZ 965A and 966 built in Ukraine, via medium-sized cars such as the Lada/Zhiguli (developed with Fiat) and Moskvitch and large cars like the GAZ Volga, through to luxury cars for senior party officials such as the Chaika and ZIL." During the period 1950–1965, while many automobiles were distributed to Communist Party elites, a small export market had developed in the Soviet Union, with most of the exported motor vehicles went to Soviet bloc countries (Siegelbaum 2008:224). Whereas in 1946, the Soviet Union manufactured 6,289 automobiles, 94,572 trucks, and 1,310 buses, in 1964, it manufactured 185,159 automobiles, 385,006 trucks, and 32,919 buses (Siegelbaum 2008:219). Despite manufacture of some automobiles, initially the Soviet Union and other Soviet-bloc countries "rejected private ownership of cars as an alien concept while emphasising collective uses of transportation as an alternative to capitalist private ownership" but after the 1950s and 1960s, Communist Party elites redefined the automobile as a "veritable socialist item" along with other consumer items, such the modern kitchen (Fava and Gatejel 2017: 12). Several Eastern European countries, including Poland, Yugoslavia, Romania, and even the German Democratic Republic, developed technical support arrangements with Western European motor vehicle manufacturers. An exception to this pattern was Czechoslovakia which developed an independent national automobile industry during the 1960s and 1970s (Vilimek and Fava 2017). While the density of automobiles per 1,000 capita had significantly increased in all the republics of the Soviet Union between 1977 and 1985 in what ultimately proved to be the last decades or so of that country, there continued to be significant variation among the republics in terms of the density of automobile ownership (Siegelbaum 2008:240).

Motorization historically lagged in the Soviet Union behind the advanced capitalist societies, in part due to the underdevelopment of roads in a vast country that was more easily traversed by railroad lines, particularly the Trans-Siberian Railway. Nevertheless, in its last days, the Soviet Union in 1990 manufactured some 1.2 million automobiles and 780,000 trucks (Siegelbaum 2008:253). In its last days, the USSR produced some 1.2 million automobiles and 780,000 trucks (Siegelbaum 2008:253). Siegelbaum (2008:256) reports: "Car ownership in Russia, which stood at 75 per thousand inhabitants in 1993, reached 150 ten years later. The total number of cars was approximately 11 million in 1992, surpassed 20 million by 2000, and reached 25 million in 2003."

The private automobile density was highly variable among the various Eastern-bloc countries, most of them having a higher density than the Soviet Union. For instance, 1989, whereas the private automobile density in the Soviet Union was a mere 43.4 automobiles per 1,000 people, in Bulgaria it was 137.0, in Czechoslovakia 200.0, in the German Democratic Republic 232.6, in Hungary 163.9, in Poland 126.6, and in Yugoslavia 135.1 (Siegelbaum 2011:8). This higher density undoubtedly was partly a by-product of an already-existent culture of automobility prior to the formation of a socialist-oriented political system in Eastern European in the wake of World War II. For example, in the case of Czechoslovakia, "starting from the first half of the twentieth century, the automobile, which was destined for exportation and racing, had become one of the symbolic manufactured products of the Bohemian 'industrial tradition'" (Fava 2011: 17).

Whereas the Soviet Union and other Eastern-bloc countries increasingly came to embrace the culture of automobility, Albania, a socialist-oriented or post-revolutionary country which under Hoxha distanced itself from the Soviet-bloc for the most part rejected it. In 1969 Albania, which had over two million people, had some 2,700 passenger vehicles and 7,700 trucks and buses (Dalakoglou 2012:574). Private automobiles were forbidden until March 1991. Ironically, the Albanian state required its citizens to work on the construction of roadways as an exercise promoting purported socialist modernity that in essence integrated the isolated mountainous districts into a unified political entity.

Since the collapse of the Soviet bloc, automobility has taken the region by storm, to such an extent that "cars now terrorize pedestrians from Estonia to Bulgaria," accompanied by the massive road construction and the abandonment of intercity rail lines (Crawford 2002:89). To some degree, various automobiles companies, some domestic and others foreign, have come to dominate the automobile market in various countries. Skoda Auto dominates the market in the Czech Republic and Slovakia, Renault in Slovenia, Renault's subsidiary Automobile Dacia in Romania, Fiat in Poland, and Suzuki in Hungary (Business Monitor International 2009:15). Because Russian automobiles cost about one-fifth of what foreign automobiles cost, they continue to be understandably popular in Russia. Conversely, the Russian government is encouraging joint ventures between Russian automobile and foreign automobile manufacturers, indicated by that fact that in February Prime Minister Vladimir Putin "said that he would back a new joint venture between Fiat and Russia's Sollers in the central Russian car producing city of Naberezhnye Chelny" (Lewis 2010:47). Nissan and other foreign automobile companies have manufacturing plants in St. Petersburg. In Bucharest, Romania, the increase in private automobiles has resulted in parking shortages as well as self-appointed parking attendants helping drivers to find a parking spot for a fee, a new source of income helping marginal individuals and families to survive in an economically ravaged society (Chelcea and Iancu 2015).

Production and Consumption of Motor Vehicles in Developing Societies

In recent decades, automobile firms have been searching for new markets in the Third World and, with the collapse of the Soviet bloc, in Eastern Europe. David Harvey (2014:58) argues that with "relative saturation of the market for new automobiles" in developed capitalist societies, the "auto industry now looks, therefore, upon those unsaturated markets in China, India, Latin America, and the deliberately 'under-urbanized' world of the former Soviet bloc as its primary realm of future accumulation." Furthermore, as Shaefer et al. (2009:3) observe, "[s]ince about 80 percent of the world population lives in the developing countries, the largest wave of motorization is yet to come." According to Dicken,

> [I]n the Americas, both Canada and Mexico are tightly enmeshed with the US automobile industry . . . while Brazil remains the major automobile production centre in Latin America. The most striking new development of recent years has been the sudden emergence of South Korea as an important producer. As recently as the early 1980s, Korea was producing only 20,000 automobiles. In 2000 Korean output was 2.4 *million* (6 percent of the world total). Thailand has evolved into the "car capital" of Southeast Asia with many foreign companies having manufacturing facilities there. (Dicken 2003:359)

These operations are dwarfed by a burgeoning car manufacturing sector in China and India. China did not permit private motor vehicle ownership until 1984; however, in 1994 the Chinese state designated the automobile industry as a key pillar of economic growth and encouraged citizens to purchase private motor vehicles (Leung 2010). The Chinese car industry consists of state companies as well as joint operations between these companies and foreign companies, including Volkswagen, Toyota, Nissan, Honda, Hyundai, and General Motors (Dicken 2003:396–397). China does not allow foreign multinational corporations to assemble motor vehicles or manufacture vehicle component parts but requires them to enter joint partnerships.

The most dramatic instance in the growth of motor vehicles in the developing world is China. Jared Diamond (2005:362) reports that the "number of motor vehicles (most trucks and buses) increased 15-fold between 1980 and 2001, cars 130-fold." In 2002, 1.2 million passenger vehicles were sold in China, and the following year this figure had grown by 20 percent (Rutledge 2006:135). In addition to exclusively state-operated motor vehicle corporations in China, in 2008 there were seven major passenger automobile joint ventures in China (Luethje 2014:543). According to Luethje (2014:544), "[t]he traditional regions of motor vehicle production in China have been Changchun in the northeast and Hebei in central China. Newer key locations for the Chinese motor vehicle industry include Shanghai and the great Yangzi River Delta, the Beijing-Tianjin corridor, and the Pearl River Delta in southern China."

Whereas vehicle sales dropped 50 percent in the United States and 40 percent in other countries overall in 2009 during the Global Financial Crisis, China overtook the US during that year as the largest domestic market for cars in the world and became General Motor's largest market (Achoff 2011:134). Vehicle production in China declined during the last eight months of 2008 but then "rose steeply enough to resume or even exceed the pre-2008 trend" (Gilbert and Perl 2010:303). In 2010 Chinese factories manufactured 13.9 million automobiles in 2010 and this figure soared to 18.1 million in 2013 (International Organization of Motor Vehicle Manufacturers 2014). Private motor vehicle ownership in China hit 73 million in 2011 (Newman and Kenworthy 2015:82).

The Chinese motor vehicle industry employs some 1.7 million workers (Klare 2008:70). Given that only five percent of Chinese households owned a car in 2009, the potential for an increase in the number of cars in China is huge (Montgomery 2010:8). Various studies predict that China will have approximately 200 million cars by 2020, nearly 400 million cars by 2030, and 700 million cars on the road by 2050 (Montgomery 2010:37; Bongardt et al. 2013:20). Various factors may curtail these projections, such as oil depletion, socioeconomic, ecological, and climatic crises. Nevertheless, the Chinese state has provided massive support for creating the infrastructure that the culture of automobility requires. Calthorpe reports:

> In the last five years, China has built more than 30,000 kilometers of expressways, finishing the construction of 12 national highways a whopping 13 years ahead of schedule and at a pace four times faster than the United States built its interstate highway system. Over the last decade, Shanghai alone has built some 2,400 kilometers of road, the equivalent of three Manhattans. . . . China already has passed the United States as the world's largest automobile market, and, by 2025, the country will need to pave up an estimated 5 billion square meters of road just to keep moving. (Calthorpe 2016:94)

To accommodate the explosive increase in motor vehicles China is constructing more highways, ring roads, and parking lots, much as the United States did in an earlier era. Despite such efforts, traffic in Beijing for example comes to a grinding halt on a regular basis. In August 2010 a highway on the periphery of Beijing experienced a 96 kilometer or 60-mile traffic jam that lasted eleven days (Calthorpe 2016:95). Such scenarios have prompted Beijing, Guangzhou, and Shanghai to implement a lottery to restrict the number of motor vehicle registrations.

Whereas in 1951 India had an estimated 300,000 cars, this number had increased to about 85 million by 2005 (Shiva 2008:52). In short, more vehicle production has been a key component of Indian industrial development since the late colonial period (Tetzlaff 2017). Indian cities are rapidly being motorized on various fronts. Reddy and Balachandra observe:

> Motorcycles in particular, as well as cars, are burgeoning as major forms of personal mobility, while walking and bicycling, once very prominent in cities, have taken a back seat. Even though public transport offers a competitive service, its market share is constantly declining. In fact, there is a transition from public transit-orientated mobility towards private transport. (Reddy and Balachandra 2012: 152)

Tata launched the mini-car Nano in India in 2008 as an alleged "people's car." It plans to market the Nano in Southeast Asia, Latin America, and Africa (Montgomery 2010:37). Tata launched the minicar Nano in 2008 as an alleged "people's car" and projected producing 1 million of them per annum by 2011. However, given the aspirations of the Indian "new middle class" to own an automobile as a status symbol, Tata's marketing campaign to manufacture a people's car collapsed for the most part and the company dropped its promotion of an inexpensive "basic" Nano and in January 2014 promoted the Nano Twist as a "new smart city car" pitched at young professional urbanites (Nielsen and Wilhite 2015:382; Notar 2017). General Motors, Honda, Volkswagen, and other companies plan to build new factories in India and Fiat, Nissan, and Renault are forming partnerships with Indian car manufacturers (Shiva 2008:50). In 2005, Bangalore, a teeming Indian metropolis of some 8 million inhabitants, had 16,710 private two-wheelers, 3,510 private automobiles, and 810 auto rickshaws and cabs (Rahul and Verma 2014:108). The Bangalore Metropolitan Transport Corporation, a public sector company, constitutes the major public operator in the city but the percentage of ridership had declined as the percentages for private two-wheelers and private automobiles for the most part have risen between 1985 and 2005. Bangalore, in contrast to Mumbai, lacks a regional railway system. Although the number of people using auto rickshaws and cabs has risen during this period, largely due to a growing population, the percentage of ridership declined between 1985 and 2005.

Although Mumbai, a megacity with a population of some 19 million people, has a regional railway system, its motor vehicle mix is projected to rise along with population increase and a growing number of affluent residents who will adopt private motor vehicles. For example, whereas in 1998 the city had 145,461 petrol automobiles and 36,365 diesel automobiles, one study projected that in 2020 it would have 833,383 petrol automobiles and 133,556 diesel automobiles (Yedla, Shrestha, and Anandarajah 2005:249). In terms of motor vehicle utilization, annual travel by private passenger cars (passenger kilometers per capita) in 1990 stood at 19,004 in Houston; 16,686 in Los Angeles; 9,417 in Sydney; 4,482 in Paris; 3,175 in Tokyo; 6,299 in Kuala Lumpur; 4,634 in Bangkok; 2,464 in Seoul; and 1,546 in Jakarta (Newman and Kenworthy 1999:84).

Most of the world's largest automobile manufacturers carry out over 40 percent of their production activity outside of the country where they have their headquarters (Paterson 2007:98). In many ways, the private motor vehicle is the ultimate commodity which is in turn powered by another commodity, namely oil. As a source of profits, the financing of automobiles has overtaken manufacturing of them as a source of profits (Lutz 2014:240).

Powered two-wheel vehicles (PTWs) include motorcycles, but also motor scooters and mopeds. In many developing countries PTWs "provide affordable everyday transport for families and workers and are used for commercial and public service activities" (Pinch and Reimer 2012:441). In 2008, of the total global sales of 38.5 PTWs, over 85 percent occurred in Asian developing societies, particularly China, India, Indonesia, Thailand, and Taiwan (Pinch and Reimer 2012:441). Conversely, in large part due to the impact of the global financial crisis on Spain, between 2004 and 2012 Barcelona experienced a slight decline in automobile ownership but a marked increase in motorcycle ownership (Marquet and Miralles 2016).

There are over 301 million PTWs in the world, with 94 percent of them in developing countries. In 2014, there were 382,160 motor vehicles in Europe, with a motorization rate of 464 per 1,000 inhabitants; 316,630 motor vehicles in the NAFTA countries (United States, Canada, and Mexico), with a motorization rate of 661 per 1,000 inhabitants; 87,129 motor vehicles in Central and South America, with a motorization rate of 176 per 1,000 inhabitants; 407,874 motor vehicles in Asia/Oceania/Middle East, with a motorization rate of 100 per 1,000 inhabitants; and 42,511 motor vehicles in Africa, with a motorization rate of 44 per 1,000 inhabitants (OICA 2014). Globally in 2014 there were 1,236,273 motor vehicles, with a motorization rate of 180 per 1,000 inhabitants (OICA 2014).

Tractors constitute a specialized type of motor vehicle, one that is not used generally for transportation per se, but in agriculture. Henry Ford pioneered the mass production of tractors and there are currently about 30 million tractors in use around the world (Schwaegerl 2014:91).

The Nexus between the Motor Vehicle Complex and the State

The role of the state in advanced capitalist societies has been to resolve the contradictions that develop in a market economy and to reduce social conflicts that may threaten the stability of the social system. The state must be responsive both to the demands of the capitalist class and the public, particularly labor. Although the state must cater to the latter to some extent, it never questions the logic of the corporate economy with its imperative for continual profit-making and economic growth. Consequently, when the state promotes changes in public policy, including those related to various aspects of motor vehicle production and highway construction, they tend to be congruent with the interests of the corporate sector.

While corporations often assert that they favor minimal government involvement, the collusion between the motor vehicle industry and governments around the world constitutes yet another manifestation of *state capitalism* par excellence. As Dicken observes,

> [T]he state has played an extremely important role in [the automobile industry's] evolution. In particular, trade barriers have exerted an extremely important influence in both developing and developed economies. At the same time, national governments have struggled to outbid one another . . . to secure the large manufacturing industries. . . . The giant TNCs of the industry have developed consummate skills in playing governments off against each other. (Dicken 2003:353)

Developed Democratic Capitalist Societies

The nexus between the motor vehicle complex and the US is perhaps illustrated by Charles Wilson's assertion when he was the chairperson of GM that "What's good for America is good for GM, and vice versa" (quoted in Quinn 1988:47). In the United States the nexus between the motor vehicle complex and various governments extends back to the early twentieth century. American automobile clubs and automobile manufacturing lobbies in the guise of the Good Roads movement urged county governments to build exurban roadways (Black 2006:117). In 1916 Congress passed the first of numerous federal highway funding acts, starting out with $75 million in order to build roads designed to connect rural areas to towns and cities (Black 2006:117). According to Kay, "The act called for the establishment of a highway department in every state" and required local municipalities to create highway departments to match federal funding. The American Road Builders Association and the Association for Standardizing Paving Specifications joined the campaign to build more roads.

Fordism, however, propelled the automobile from its status as a luxury commodity into a mass commodity. Due to the stock market crash of 1929, the US capitalist class found an ally in Roosevelt's New Deal, which utilized the state as a way to generate jobs, thus driving the purchase of automobiles and avoiding the problem of their overproduction.

A key component of the new corporate-state alliance was industrial activity focused on automobile production, construction, transportation, steelmaking, rubber extraction, and production of consumer goods. Foster observes:

> Road building became one of the cornerstones of the New Deal work programs during the 1930s. It was popular with politicians because its benefits could easily be spread between many congressional districts. Representatives who successfully backed such projects in their districts could claim highly visible "concrete" achievements. Equally attractive, road building was also highly labour-intensive, which meant that, in terms of jobs provided, there was a lot of bang for the taxpayers' buck. In fact, President Roosevelt's New Deal greatly enhanced "automobility" at the expense of public transport. (Foster 2003:14)

Roosevelt's vision of economic stability and democratic welfarism did not come to full fruition until after World War II—an event that did much to overcome the Depression and to propel the United States into the position of the foremost capitalist nation and the leading culture of consumption in the world. A key component of the new corporate-state alliance was industrial activity focused on automobile production, construction, transportation, steelmaking, rubber extraction, and production of consumer goods.

In the 1930s, General Motors created the National Highway Users Conference, a consortium of over 3,000 businesses, which lobbied for more and better highways, an effort that eventually culminated in 1956 with the beginnings of a federally funded Interstate Highway System (Luger 2000:12). The US Congress passed the Federal Aid Highway Act, officially called the National Interstate and Highway Defense Act. Avila reports:

> Two military men drafted the Highway Defense in its roughest form: Dwight Eisenhower, a former five-star general, and his ex-deputy Lucius Clay. During his travels in Germany, Eisenhower lauded the Autobahn as the "wisdom of broader ribbons across the land," and during his presidency he emphasized concerns in his support for a national highway system. The onset of the Cold War heightened the need for contingency plans in the event of an air attack or land invasion of the cities. (Avila 2014:26–27)

However, it is important to note the interstate highway system also served to facilitate US economic development in the post-World War II era.

The revolving-door syndrome between the US motor vehicle industry and the state is illustrated by the fact that both Charles Wilson, a former General Motors president, and Robert McNamara, a former Ford Motors Company president, both served as secretaries of defense, and that Thomas Mann, a high-ranking State Department official, became president of the Automobile Manufacturers Association. Furthermore, the highway construction companies heavily influenced party politicians in the federal, state, and local governments to pass legislation that promoted the culture of automobility.

A powerful motor vehicle-highway lobby consisting of the automobile industry, oil companies, trucking firms and road clubs historically posed a formidable barrier to the development of effective public transportation, especially in most American urban areas, but also in countries such as Australia. In the case of the United States, as Taebel and Cornehls observe,

> [C]orporate administrators and technicians, particularly those of the auto industry, have long moved free in and out of the federal government, thus blurring the line between governmental interests and those of private business concerns. Nowhere is this mutuality of interests more clearly understood or more staunchly promoted than among the leaders of the auto corporations themselves. (Taebel and Cornehls 1977:75)

The nexus between the motor vehicle industry and the state manifested itself in a somewhat different form when the US government provided Chrysler, which had run into severe financial difficulties, a bailout package in the late 1979s and the Obama did the same for the automobile industry during the global financial of 2008–2009. In the case of the Chrysler bailout, the company orchestrated a massive lobbying effort that included many of its 130,000 employees, 4,500 dealerships, and 19,000 suppliers, which highlighted the "company's economic contribution to each congressional district in the nation" (Luger 2000:101). When Douglas Fraser, the president of the United Auto Workers (UAW) proposed partial government ownership of Chrysler, the company and its political allies managed to snuff out this proposal. During the Global Financial Crisis of 2008–2009, the Obama administration with support from Democrats in the Great Lakes region supported generous bailouts for General Motors and Chrysler but these were "strongly opposed by Republicans from southern US states who viewed the bailouts as unfair subsidies to unionized D-3 [Detroit 3] firms in the Great Lakes states given that their own states are home to non-union Asian and European automakers," thus illustrating that the "state is not a unified actor and its policies are subject to often intense class and regional conflict" (Rutherford and Holmes 2014:373). A recent example of the nexus between the motor vehicle industry and the US state is the support of the National Science Foundation for development of forged nanoparticle aluminum alloys for the automobile and defense industries.

Western European governments in particular have been extensively involved in their domestic motor vehicle industries. Indeed, until recently, the French, British, and Italian states partially owned automobile manufacturing operations. After World War II, the new Labour government in the UK promoted not only the production of motor vehicles for domestic consumption but also for export. The Ministry of Supply, which had oversight over various industries, including the iron and steel and automobile industries, collaborated with not only various motor vehicle manufacturers, but also the Motor Agents Association, the British Trade Association, the Scottish Motor Trade Association, the Institute of the Motor Industry, the British Transport Vehicle Manufacturers Association, the Motor Division of the Institute of Mechanical Engineers, the Motor Factors Association, and the Society of Motor Manufacturers and Traders (McLaughlin and Maloney 1999:18–19). Indeed, in 1949 Britain was the leading exporter of automobiles in the world, with 69 percent of them going to Commonwealth countries, such as Canada, Australia, New Zealand, and India. The British Board of Trade encouraged the motor vehicle industry to "disperse its production facilities into new locations" (McLaughlin and Maloney 1999:28).

On the Continent various governments between 1945 and 1955 nationalized their motor vehicle industries (Nieuwenhuis 2014:44). France nationalized Renault and Italy nationalized Alfa Romeo, transforming it from a manufacturer of luxury sports automobiles to manufacturer of automobiles for a mass market, reportedly as a "counterweight against the very powerful and privately owned Fiat group" (Nieuwenhuis 2014:44). Both Christian Democratic and Social Democratic transportation ministers in the Federal Republic of Germany supported motorization, resulting in an increase of travel to work in automobiles and buses rather than on trains (Yago 1984:47).

The Japanese government played a central role in supporting Japanese motor vehicle companies beginning in the 1960s, enabling it to become the world's largest manufacturer and exporter by the early 1980s (Quinn 1988:17). Indeed, by the mid-1960s, Toyota and Nissan overtook Volkswagen as the major foreign-automobile exporter to the United States. Flink (1988:327) reports that "in worldwide production Japan passed Germany to rank second in 1967 and passed the United States to take first place in 1980."

Fascist Regimes

Fascist regimes have had a long track record for furthering the culture of automobility. Benito Mussolini raced a motorboat and was passionate about automobiles and speed which he shared with Filipo Tommatso Marinetti, the founder of Italian Futurism, elements which paved the way for transportation development in Italy. He reportedly built the first motorway in 1925, which connected Milan with the Alpine lakes (Mosey 2000:47). Italy had seven motorways in place by 1939. Mussolini viewed Italian motorways as a vehicle of economic development in which "industry and state would work together to accelerate national growth" (Mosey 2000:47).

Nazi Germany constitutes another example of the close ties between the automobile industry and the state. Hitler elevated Henry Ford to his pantheon of heroes, but he merely tapped into the pre-existing German fascination with *Fordismus* exemplified by the fact that in 1925 Ford's autobiography *My Life and Work* was a best seller in Berlin (Flink 1988:113). Indeed, the "automobile-oil-rubber" industrial group became increasingly influential in Germany during the 1920s, so much so that by 1929, the Reich Association of the Automobile Industry, the Association for Promoting the Hansa Cities—Frankfurt-Basel Autobahn), and other industry and trade associations had achieved significant political support for motorization and highway construction (Yago 1884:35). The Nazi assumption of state power in 1933 meant that a "community of interest was established between the automobile industry and the German fascists who pursued a national transportation policy encouraging motorization" (Yago 2003:36). For a brief period of time there was a faction within the Nazi Party which opposed motorization, but it was purged by 1934.

Hitler viewed the automobile as a device for transcending class conflict evidenced by his assertion: "I have come to the conclusion that the motorcar, instead of being a class dividing element, can be the instrument for uniting the different classes, just as it has done in America, thanks to Mr. Ford's genius" (quoted in Gartman 1994:15). His office contained a huge picture of Ford on a wall and piles of Ford's books. By late 1933 the Nazis had published numerous editions of Ford's *The International Jew* and praised him for promoting anti-Semitism around the World (Flink 1988:113). On his 75th birthday on July 30, 1938, Ford accepted from Hitler the Grand Cross of the Supreme Order of the German Eagle for promoting motoring among the masses.

In pre-Nazi Germany, as Sachs (1992:36) observes, the "luxury market, rather than mass consumption, was the midwife of the automobile, for into the 1930s the industry geared its production above all to customers who looked not to price, but first of all to beauty and performance." A British automotive writer reported in the magazine *Autocar* that the 1928 Berlin Auto Show featured 62 exhibits for automobiles, 38 of which were not German, as well as exhibits for motorcycles, trucks, and buses (Koshar 2005:126). Indeed, Germany became the foremost manufacturer of motorcycles in the world for a period. Shortly after assuming power in 1933, Hitler promised ordinary Germans that in good time they too would possess automobiles as part and parcel of his *Volksgemeinshaft* (people's society). As Paul Virilio observes,

> As soon as it takes power, the Nazi government offers the German proletariat sport and transport. No more riots, no need for much repression: to empty the streets, it's enough to promise everyone the highway. This is the "political aim" of the Volkswagen, a veritable plebiscite, since Hitler convinced 170,000 citizens to buy a VW when there still wasn't a single one available. (Virilio 1986:25)

Hitler declared at the 1934 Berlin Automobile Show that the "class-emphasizing and therefore socially differentiating character that has been attached to the automobile must be removed; the car must not remain an object of luxury but must become an object of use!" (quoted in Sachs 1992:56). Hitler systematically destroyed the hatred of automobiles held by many German small town and rural people because they disrupted the serenity of their communities. Instead, the Nazis organized the NSKK (*National Socialistiches Kraftfahr Korps* or the National Socialist Automobile Corps) at the local level which soon organized some 500,000 members and trained them to "drive over every kind of terrain, to shoot while driving, etc." (Virilio 1986:25). For the most part, automobiles became a rage, often for recreational purposes, among middle-class Germans rather than working-class Germans (Mom 2015:336).

Hitler regularly read motoring magazines, followed auto racing, and viewed himself as an automotive engineering expert, despite that he never learned to drive (Patton 2002:8–9). He particularly loved fast Mercedes sedans and purchased one in 1923 from Jacob Werlin, an automobile salesman who eventually became a member of the Mercedes-Benz board of directors and Hitler's chief advisor on automotive affairs (Flink 1988:261). In 1933 Hitler unveiled his strategy to rejuvenate the declining German auto industry at the International Automobile and Motor Cycle Exhibition in Berlin a few days after assuming power by announcing plans for the development of the Volkswagen and the autobahn. He opened the first autobahn between Cologne and Bonn in 1932 and orchestrated the construction of 2,107 kilometers of autobahns in the first decade of his rule (Mosey 2000:47). However, the autobahns during the Nazi era tended to be relatively free of automobiles, largely because much of the Nazi-sponsoring manufacturing effort focused on military hardware and arms, including military motor vehicles. However, tour buses began to compete with the railway in catering to the tourist trade during the period of 1933–1939 (Vahrenkamp 27/2).

Indeed, Hitler's motorization campaign became the central plank in his program to eliminate unemployment in Germany. He also lifted taxes on new cars as a mechanism for promoting the culture of automobility (Mom 2015:291). Flink (1988:262) reports: "Production of motor vehicles increased from only 52,088 units in 1932 to 342,169 units in 1938, the last peacetime year. Between 1925 and 1930 only 89,000 commercial vehicles had been produced, versus 265,000 between 1933 and 1938. Tractors in use increased from 25,000 in 1929 to 82,000 in 1939."

Alfred P. Sloan, the president of General Motors, like Ford, exhibited far right-wing tendencies, which legitimized his company's operation in Nazi Germany. Under Sloan's leadership, GM added German automobile firm Adam Opel to its portfolio.

> [H]e was certainly happy to appease the Nazis after 1933, acquiescing to the Nazification of Opel's plants and management, with the anodyne comment that "politics should not be considered the business of the management of GM," Sloan said nothing as the Nazis purged all the Jewish employees from his German factories and dealerships, and did not interfere as Opel's new Brandenburg plant was forcibly converted to make military trucks. When the European war broke out in 1939, he famously declared that "we are too big to be inconvenienced by these pitiful international squabbles." (Parissien 2013:90)

James Mooney, the President of GM Overseas Corporation, participated in the "America First Movement to keep the United States out of the war" (Yago 2003:39). Albert Speer, the Nazi regime's minister of production, credited GM technical assistance for its ability to invade Poland in 1939.

Although Sloan opted not to intervene in Nazi politics, he adopted a double-standard when it came to US politics, as evidenced by the fact that he bankrolled the far-right American Liberty League which mounted a vehement campaign against Franklin D. Roosevelt's 1936 bid for the presidency. The British Road Federation was so taken by the Nazi's autobahn system that they invited in 1934 the British Ministry of Transport officials to investigate it, but they did not do so (Mosley 2000:47). Not so easily deterred, the federation in 1937 sent a delegation of 255 people, including 58 Members of Parliament and 54 county surveyors, to inspect the autobahn system. The Volkswagen plant produced military jeeps and Daimler-Benz and BMW produced aircraft engines, tanks, and armored trucks (Wolf 1996). Opel, owned by General Motors, and Ford also manufactured products for the Nazi war machine (leading to US government payments to those companies for damages as a result of US bombing of Germany during the war).

Socialist-Oriented or Post-Revolutionary Regimes

Political leaders in post-revolutionary or socialist-oriented societies often embraced automobilization. V.I. Lenin addressed a crowd from atop an armored automobile after returning from exile in Switzerland at Finland Station on April 3, 1917 (Siegelbaum 2008:181). The Bolsheviks expropriated the 46 automobiles from the Czar's personal garage after the Revolution which included "several Delaunany-Belleville limousines, a few Mercedes, a Daimler or two, a six-cylinder Renault, a Rolls-Royce Silver Ghost, an a Turcat-Mery—along with its staff of expert mechanics and drivers" (Siegelbaum 2008:181). Lenin often rode in an open Delaununay-Belleville on his trips to meetings outside the Kremlin or country outings.

Soviet leaders in the 1920s greatly admired Fordism and built cities guided by modernist principles which included tall buildings, broad streets, and free-flowing traffic, possibly because few people actually owned cars. V.I. Lenin reportedly owned a Rolls-Royce Silver Ghost. Some 25,000 Fordon tractors were shipped to the Soviet Union between 1920 and 1927. Indeed, Stalin referred to Ford as "one of the world's greatest industrialists" (Wik 1972:4). The Soviet Union adopted Fordist production methods in its factories. Flink (1988:112–113) reports: "Progress in adopting them was chronicled in *Pravda*, and in workers' processions Ford's name was emblazoned on banners emblematic of a new industrial era. Translations of *My Life and Work* were widely read and used as texts in the universities."

Stalin accepted a Mercedes-Benz as a gift from Hitler when the two were on better terms, and Brezhnev proudly drove a Cadillac Eldorado that Nixon had given him in 1972 (Marsh and Collett 1986:23). Khrushchev proposed that "socialist motorization take the form of collectively owned vehicles"

(Ladd 2008:62). His legacy in terms of automobiles was rather contradictory in that "privileges for the elite co-existed with more open access to cars, in the form of rentals and taxis" with private ownership of cars being on the rise but still "ideologically off-limits" (Gatejel 2016: 130).

In the aftermath of the Chinese Revolution, Mao Zedong during his trip to Moscow between December 1949 and February 1950 visited the Stalin Auto Works, primarily a huge truck factory (Seow 2014). Of the 50 heavy industrial construction projects that the Soviet Union had committed itself to in the new People's Republic of China, one of which was a motor vehicle factory, namely the First Auto Works which was completed on October 15, 1956, in Changchun. It was designed to manufacture some 30,000 medium trucks a year that would play a crucial role in the economic development of the PRC. The "Liberation truck" was a copy of the Soviet ZIS-150, which was itself a copy of an International Harvester model, and was well-suited for travel in high elevation regions and dirt roads. First Auto Works also manufactured two passenger automobiles, "Eastern Wind" and "Red Flag," both of which were first produced in 1958. The latter was used primarily to chauffeur Communist Party elites around China. By 1960, China had sixteen motor vehicle factories, including ones at Beijing, Nanjing, Shenyang, Shanghai, and Jinan (Seow 2014:152). In recent decades the Chinese state has promoted a massive automobile manufacturing industry, either in the form of state companies or joint-venture companies and has adopted "favourable tax policies and government incentives that encourage the purchase of automobiles and motorcycles" (Cervero 2013:8). Imitating the Chinese model of modernization, Vietnam has embarked upon a massive program of motor vehicle production, one that has been shifting quickly in the past two decades from motor bikes to automobiles (Hansen 2017). However, the former much more agilely weave their way through the crowded streets of Hanoi, Ho Chi Minh City, and other Vietnamese cities than are the latter.

Motor Vehicles and Militarism

A special component of the nexus between the motor vehicle industry and the state is the close connection at different times and places between militarism and motor vehicles. This is a connection that historians of the automobile have generally touched upon tangentially, yet it is a story which needs to be told. Indeed Kurt Moeser (2003:239) maintains that during the "formative decades of automobilism the use of automobiles and aeroplanes in the context of races and other technologically oriented mass spectator sports played a significant role in generating a collective mood that anticipated and prepared individuals and societies for a European war," namely World War I.

The earliest military experimentation with motor vehicles appears to have occurred in 1897 when at the French army's annual maneuvers they were utilized (Flink 1988:73). The British army first used motor vehicles in the colonial service and shortly thereafter in the Boer War in 1899–1902. Over the course of the twentieth century and into the twenty-first century military forces have relied on a wide array of motor vehicles, ranging from automobiles, jeeps, and Humvees to trucks to tanks. Reportedly in July 1914 the French Army owned 220 motor vehicles, including 91 trucks and 50 tractors for pulling artillery (Bardou et al. 1982:23). Flink (1988:73–74) observes: "By the outbreak of World War I, officers rode in staff cars and couriers drove motorcycles in European armies: the artillery tractor had been developed in France, the tank in Britain, and the armored car in Britain and Germany."

In the case of France, "65,592 trucks and cars and some 3,200 tanks were made for the military during the war, versus only about 2,500 motor vehicles for the private sector" (Flink 1988:75). German motor vehicle production increased profoundly during World War I, but not sufficiently to meet its military needs. Although the US Army Signal Corps purchased a few light trucks and an automobile ambulance in 1906, the US military was rather slow in adopting motor vehicles on a large-scale basis. The United States entered World War I belatedly in 1917 but adopted the Model T Ford in large numbers during the war effort. In contrast to the UK which used almost 100,000 trucks and automobiles, the US used some 50,000 motor vehicles (Angus 2016:131).

In the case of Japan, its army began to show interest in adopting motor vehicles in 1907. Flink reports:

[B]eginning with the Military Motor Vehicle Subsidy Act of 1918, the Japanese government offered subsidies to Japanese-owned manufacturers to produce motor vehicles suitable for military use. Yet few were produced before 1936, because the well-established *zaibitsu* (family-owned business groups), to whom the military were hostile, remained reluctant to risk capital to compete with Ford and GM. Consequently, the military established alliances with newer business groups—principally Toyota and Nissan in the case of motor vehicles. (Flink 1988:271)

The German High Command under the Nazi regime knowing that the Germany had suffered militarily during World War I due to its low level of motorization "pushed for the production of military vehicles of all types—trucks, motorcycles, and later, tanks" (Yago 1984:41). As the people's car, the commercial Volkswagen or the KdF (Kraft durch Freude or Strength Through Joy) project essentially was put on hold during World War II and transformed into a military project. Flink reports: "During the war versions of the KdF-Wagen were used by the German military as general-purpose vehicles. The most important were the Kommandereur (command car), the Kuebelwagen (literally, "bucket car"—the counterpart of the American jeep), and the amphibious Schwimmwagen."

Ironically, whereas the Allied war production resulted in some 660,000 jeeps, the Nazi state only manufactured 50,435 Kuebelwagens. Ferdinand Porsche, the Austrian engineer who Hitler had commissioned to design the Volkswagen, also designed the Leopard tank and his design team consulted on the development of the V-1 buzz bomb or *Vergeltungswaffe-eins* (Revenge Weapon One).

The Nazi invasion of the Netherlands, Belgium, and France in 1940 prompted the Roosevelt administration to create the National Defense Advisory Commission on May 29, 1940 which developed a voluntary plan to shift US industry toward military production, with Ralph Budd, the president of the Association of American Railroads, in charge of the agency's transportation division (Gilbert and Perl 2010:25). Because the agency did not make much progress in its mission, the Roosevelt administration established the Office of Production Management (OPM) to mobilize military manufacturing and the Office of Price Administration and Civilian Supply (OPACS) to "regulate non-military production at a level that could support optimal military output" (Gilbert and Perl 2010:26). The Japanese attack on Pearl Harbor on December 7, 1941, prompted the United States to enter World War II. The US state ordered the automobile industry to convert its manufacturing operations into producing a wide array of military hardware, not only motor vehicles and tanks but also aircraft, parts for aircraft engines, and artillery ammunition (Doyle 2000:1–2). This process, however, had started even before the Japanese attack on Pearl Harbor. As early as November 1, 1940, the Ford Motor Company had signed a contract to manufacture Pratt and Whitney aircraft engines for the US Air Force (Flink 1988:274). Packard manufactured Rolls-Royce engines for the Royal Air Force. Shortly after the attack on Pearl Harbor, the Automobile Manufacturers Association sponsored the creation of the Automotive Council for War Production.

Ford and Willy's Overland manufactured over 660,000 Willys jeeps between 1939 and 1944. In contrast, production of commercial motor vehicles declined to an almost negligible figure, 143 in 1943 for example (Gilbert and Perl 2010:27). The Roosevelt administration also imposed gasoline rationing beginning in May 1942 so that fuel could be diverted to the war effort, but this campaign only applied to 17 eastern states, not to the rest of the country, including the West Coast. According to Flink (1988:276), "In addition to turning out several million motor vehicles of various types, before the war ended the American automotive industry had produced for the military 4,131,000 engines, including 450,000 aircraft and 170,0000 marine engines: 5,947,000 guns; and 27,000 competed aircraft." Angus (2016:140) reports that US motor vehicle companies earned "some $29 billion to produce over three million jeeps and tracks, as well as airplane engines, tanks, armoured cars, machine guns, and bombs." After World War II "surplus military jeeps inaugurated a market for off-road recreational vehicles" in the United States (Flink 1988:211).

Japan, which became one of the Axis powers during World War II along with Germany and Italy, under the direction of its army shifted production away from motor vehicles to aircraft production (Flink 1988:272). After its bombing attack on Pearl Harbor on December 7, 1941, in contrast to the United States, Japan lacked enough steel to manufacture both airplanes and motor vehicles.

After World War II, the onset of the Cold War provided the primary impetus for the development of the interstate highway system. As McNeill and Engelke (2014:156) observe, "[l]ike most acts of government, this decision had many motives behind it, but prominent among them was military preparedness in expectation of war with the USSR."

Despite the end of the Cold War due to the collapse of the Soviet Union in 1991, the US military continues to possess numerous motor vehicles of various sorts. Sanders reports:

> According to its own figures, the DoD inventory of fleet vehicles worldwide—including passenger cars, buses, light trucks, and so on—totals 187,493, 13 percent of which it houses overseas. The Army and Marine Corps own and operate their own tactical wheeled vehicles, such as 140,000 High-Mobility Multipurpose Wheeled Vehicles (the HMMWV, or HUMVEE). The Army also operates over 4,000 combat vehicles and several hundred fixed wing aircraft. (Sanders 2009:47)

The May 2007 issue of *The Energy Bulletin* reported that the US military is the single largest consumer of energy world-wide and the Department of Defense's total energy consumption in fiscal year 2006 was one quadrillion (1,000,000,000,000,0000) BTU (cited in Sanders 2009:49).

THE PRIVATE MOTOR VEHICLE AS THE ULTIMATE CAPITALIST COMMODITY

While automobiles are often associated with cities, at least in the United States "rural Americans initially adopted the car as their principal means of transportation more than did their urban compatriots" (Ling 1990:7). Ling (1990:7) maintains that Progressive reforms utilized the automobile as a mechanism for integrating rural Americans into an increasingly capitalist industrial social order. Nevertheless, the automobile industry quickly viewed urbanites as targets for their products. Whereas in 1927 automobile consumption stood at 54 percent of families in cities over 100,000 per capita, it stood at over 60 percent of families in towns under 1,000 people (Ling 1990:14). Furthermore, trucks or truck farming ensured quicker delivery of agricultural products to urban markets. According to Ling (1990:33), the incorporation of rural areas facilitated by automobility "extended the market, yet at the same time it quickened the pace of distribution and sales."

The advent of large-scale motor vehicle production required mass consumer consumption, which required that not only affluent people purchase automobiles but also ordinary wage-earning working-class people. Henry Ford systematically embarked upon a campaign to socialize his and other workers into core American values which included the purchase of one of his automobiles at a purportedly affordable price. People were induced to sacrifice other needs and desires in order save up the money to purchase a Model T Ford. In the case of Ford's workers, the rising of wages to $5/day resulted in a rigid discipline mechanism.

> The $5-a-day rate was about half pay and half bonus. The bonus came with character requirements and was enforced by the Socialization Organization. This was a committee that would visit the employees' homes to ensure that they were doing things the "American way." They were supposed to avoid social ills such as gambling and drinking. They were to learn English, and many (primarily the recent immigrants) had to attend classes to become "Americanized." Women were not eligible for the bonus unless they were single and supporting the family. Also, men were not eligible if their wives worked outside the home (Worstall 2012).

Ford also hired African Americans because they constituted a cheap labor pool and by 1926 this number had risen to some 10,000 (Ling 1990:76). In contrast to European migrants who often opted to reside close to River Rouge industrial complex in places such as Delray Village, the east side of Detroit, and Hamtramck, most black workers had to commute a long distance to the plant from their ghetto in the inner-city. Indeed, their ridership on Detroit's trolley cars prompted the "switch to auto commuting by racially prejudiced whites" (Ling 1990:77).

The reality that North Americans love their cars is captured by James J. Flink's book *The Car Culture*. He observes, "During the 1920s automobility became the backbone of a new consumer-goods-oriented society and economy that has persisted into the present" (Flink 1973:40). Automobile ownership in the United States increased from a mere 8,000 in 1900, to 912,000 in 1912, to 3.4 million in 1930, to 32 million in 1940 (Parissien 2013:97). While the Depression may have prevented some Americans from purchasing and owning cars, it did not prevent others from doing so. The automobile boom in the United States during the 1920s was spurred on by provision installment plans by which consumers could pay off their car purchase over a period of a few years (Flink 1988:189). Even poor Americans fell under the spell of the culture of automobility, even during the height of the Great Depression.

Particularly in the wake of World War II, it had become commonplace for the vast majority of American families to own an automobile, a pattern that Ford had helped set in motion. Private motor vehicles perhaps more than any other consumer item exemplify the culture of consumption par excellence. During the Cold War era of the 1950s and early 1960s, General Motors urged patriotic Americans to "see the USA in your Chevrolet." Such advertisements on the part of the automobile industry served to seduce North Americans away from what had once been a relatively well-developed mass transportation system, which included passenger trains, numerous intercity bus lines, and extensive urban and interurban trolley or tram lines. The automobile came to symbolize the affluence that many young working-class Americans came to enjoy compared to the socioeconomic circumstances of either their parents or themselves in earlier times (Moorhouse 1983). As Gartman (1994:41) observes, in their efforts to contain labor agitation, corporate capitalists found a solution in "culturally integrating the working masses into the realm of privatistic consumption pioneered by the bourgeoisie," with the automobile serving as a significant instrument in this process.

While automobile sales slumped during the Depression and World War II, automobile sales exploded after the war. "In 1955 Americans bought a record 7.9 million cars, 19 percent more than the previous best sales year in 1950" (McCarthy 2007:101). While women drove automobiles prior to World War II, the increasing entry of women into the work force resulted by the 1960s in the advent of the two-car family since the husband and wife worked in separate work sites, often separated by great distances from each other (Walsh 2008:380). Even those women who did not have paid jobs, particularly ones living in the suburbs, needed a car to go shopping or chauffeur children around to various activities.

In describing the economic situation in US society during the 1970s, Paul Sweezy (1973:7) contended that the "private interests which cluster around and are directly or indirectly dependent upon the automobile for their prosperity are quantitatively far more numerous and wealthy than those similarly related to any other commodity or complex of commodities in the US economy." Automobile advertisements frequently have promised and continue to promise their target populations that they will achieve power, prestige, sexual prowess, and desirability if they choose to become the proud owners of a highly individualized form of transportation. Automobile companies in their advertisements often depict cars in natural, often remote settings, such as mountain tops, canyon lands, and secluded beaches. As Low et al. (2005:234) observe, "Nature, like sex, sells cars! And yet nothing despoils nature like cars!"

In 1990, Americans spent 31.3 percent of their incomes on housing and 18.1 percent of their income on motor vehicles (Freund and Martin 1993:16). American households routinely have a car for both husband and wife and often one or more cars for adolescent children, who in many states may obtain a driving license at age 16. At least in the United States, by 2000 the number of female drivers was virtually the same as male drivers (Walsh 2008:380). In keeping with the assertion made by George W. Bush that "Americans are addicted to oil," Foster (2003:124) reports that the "average American driver consumes five time as much gasoline per capita as his European counterpart, and *ten* times the amount used by typical Japanese drivers."

In both developed and developing countries, the car has come to be viewed, for many people, as a necessity rather than a luxury, one which often traps them in a complex web that functions, as anthropologist Daniel Miller (2001:2) describes, as a "villain that has separated us from the world and threatens to take over as we come to serve it more than it serves us." The automobile functions as a form of technological hegemony in that, as Andre Gorz, observes:

> mass motoring effects an absolute triumph of bourgeois ideology on the level of daily life. It gives and supports in everyone the illusion that each individual can seek his or her own benefit at the expense of everyone else. (quoted in Seiler 2008:146)

At the beginning of the twenty-first century, motor vehicle ownership stood at 740 per 1,000 people in the United States, 640 per 1,000 in Japan, 610 per 1,000 in Australia, 206 per 1,000 in South Korea, 190 per 1,000 in Brazil, 58 per 1,000 in Indonesia, 31 per 1,000 in India, and 21 per 1,000 in China (Vasconcellos 2001:15). While many working-class people, particularly in developed societies have become automobile owners, automobile ownership around the world constitutes a marker of social status, as well as social exclusion. For example, in the mid-1990s, in the UK, whereas 96 percent of the top 10 percent income earners own automobiles, only 25 percent of the bottom ten percent income earners own automobiles (Blow and Crawford 1997).

Although Europeans are not as automobile dependent as their North American counterparts due to a generally better public transportation system, at the end of the twentieth century, over "60 percent of European households own[ed] one or more cars, over half of journeys to work [were] made by car and over 70 percent of all journeys [were] made by private car" (McLaughlin and Maloney 1999:3).

Slick advertisements ultimately cannot effectively stimulate car sales during periods of economic downturn. General Motors and Chrysler declared bankruptcy in the United States with GM receiving a US government bailout. The US government spent taxpayer money to rescue the automobile industry without requiring CEOs to lower their overinflated salaries, requiring the companies to guarantee jobs for auto workers, or even requiring the industry to shift much of its production from cars to public transportation infrastructure. Indeed, tax dollars were used in some cases to provide company executives with bonuses.

The Hummer probably constitutes the ultimate automobile per se, as opposed to an assortment of huge recreational or camper vehicles, in terms of personal consumer items. The Hummer is an adaptation of the Humvee, a military transport vehicle which made its debut during the Gulf War, in the early 1990s. A Humvee hood served as President George H. W. Bush's Thanksgiving dinner table during his visit to the Gulf Region and the Humvee transported comedian Bob Hope when he visited Iraq during the war to entertain US troops. One interpretation contends that the Humvee became transformed into the commercial Hummer after movie star Arnold Schwarzenegger visited the Humvee production facility in Mishawaka, Indiana, in 1991. Schwarzenegger wanted to buy one in order to make a fashion statement and include a Humvee in a forthcoming film.

While the United States undoubtedly represents the leading example of the culture of automobility, the private car constitutes a component, and often a leading component of the culture of consumption in other societies around the world, particularly developed ones, such as Germany, the United Kingdom, Italy, Japan, Canada, and Australia, but increasingly developing countries, particularly China, India, and Brazil. An A.C. Nielsen poll in 2004 found that over 60 percent of residents in each of the seven fastest-growing countries aspire to own an automobile (Sperling and Gordon 2010:4).

Also, automobile ownership has sky-rocketed in the former Soviet-bloc countries. In 1981 Prague had 284,756 automobiles for 1,183,000 people (241 per 1,000 capita), a relatively high number and percentage for Soviet-bloc countries (Newman and Kenworthy 2015: 99). By 2005, Prague exceeded 500 automobiles per 1,000 capita, a figure exceeding the ownership of the three major Canadian cities of Montreal, Toronto, and Vancouver and roughly on par with Dusseldorf and Stuttgart in Germany. This increase occurred despite Prague having an excellent public transit system.

CONCLUSION

This chapter examines the history of the production and consumption of motor vehicles historically in advanced capitalist societies, the USSR and Eastern bloc countries, and in developing societies. It also discusses the nexus between the motor vehicle and the state in developed democratic capitalist societies, fascist societies, and socialist-oriented societies, which includes a discussion of the link between motor vehicles and militarism. I also examine that motor vehicle as the ultimate capitalist commodity, a theme which we see elaborated upon in chapter 2.

Although the first automobiles appeared in the late nineteenth century, automobility as a form of transportation became hegemonic during the twentieth century and continues to be so in the early twenty-first century. Low and O'Connor astutely observe:

> The form of transport that most perfectly matched the individualistic ideal of the free market was the private motor vehicle. Private cars greatly increased the choice of where to live and work, and freed labour from the tyranny of proximity to workplaces. As a mode of travel the car combined speed, flexibility, and control over personal space. . . . For firms, private trucks could deliver goods 'just in time' to meet demand wherever, whenever they were wanted, unlimited by timetabled services at railheads or ports. (Low and O'Connor 2013:5)

Over the course of particularly the second half of the twentieth century and continuing into the twenty-first century, motor vehicle production has become increasingly globalized, a process which has entailed the fragmentation of production to geographically dispersed assembly plants and supplier companies. Wailes et al. provide a succinct overview of the globalization of the motor vehicle industry since the 1970s:

> By the 1970s as trade barriers fell, European and especially Japanese auto companies, led by the Toyota Motor Company, were capturing new markets around the world and beginning to penetrate the US. . . . A new phase of global expansion since the late 1990s has seen a number of mergers and acquisitions and the development of strategic alliances and joint ventures. Demand in the existing 'triad' markets of Europe, North America and Japan is stagnant due to saturation, while expansion is expected in emerging markets, especially in Asia and South America. (Wailes 2008:6)

Indeed, as Pirani (2018:129) observes, motorization has profoundly impacted cities in developing countries, a process "encouraged by international agencies and governments who ignored everything urban planners had learned about damage done by car-based cities in the rich world." The growth of a domestic automobile industry has become an integral component of the governments of a growing number of developing countries, particularly China, Brazil, and India but also smaller countries such as Thailand and South Africa.

The Fordist model of motor vehicle was based upon the principles of Taylorism developed by Frederick Taylor, the inventor of time and motions studies, in 1881. However, in the era of neo-liberalism, particularly in developed societies, muscle power has been replaced by brain power. Shotwell (2015:37) observes: "Every worker is expected to be computer savvy and happily able to multi-task adroitly. New auto plants absorb fifty-seven seconds of every ambidextrous minute and the goal is sixty-one." The lean production system attempts to eliminate purportedly unnecessary time and energy expenditures. Instead of automobile workers taking a lunch break during which they gobble food and a beverage of some sort in the plant cafeteria or a nearby eating establishment, the GM plant in Lansing, Michigan, delivers food to the workers at their stations.

While automobility has provided capitalist enterprises with massive profits and access to markets, as we will see in subsequent chapters, it has taken a tremendous toll on the ecosystem and the climate, settlement patterns, social life, and both physical and mental health. Ironically, a substantial minority of the people in the world, or roughly one in thirteen, owns an automobile, although millions own motorcycles and motorbikes, thus enabling them to be part of modern motor vehicle mobility, often out of necessity due to lack of access to adequate public transportation (Register 2011:75).

Despite much discussion of the desperate need to curtail greenhouse gas emissions on the part of national, state governments, and local governments, the UN Framework Convention on Climate Change (FCCC), the European Union, and even various corporations, motor vehicle, oil, and other corporations in collaboration with states around the world continue to encourage increasing growth in automobile ownership and utilization. The state generally has assumed the responsibility of constructing roads and while taxing corporations, such as the trucking industry, for parts of these costs, most of the tax burdens for road construction falls upon ordinary taxpayers, many of whom use these roads in developed societies but only a few who use them in terms of driving upon them in developing societies. The motor vehicle-oil-highway lobby remains a pivotal component of state capitalism in various national guises, however, one which has been increasingly challenged since the 1970s. Nevertheless, as the next chapter demonstrates, the culture of automobility and other private vehicles is well entrenched around the world.

Chapter Two

Cultural Tropes Associated with Motor Vehicles

Historically motor vehicles have diffused from their origins in Europe and the United States to virtually every country in the world, although to varying degrees and forms. However, their utilization is mediated by cultural traditions in the many societies in which they now operate. In the introduction of his anthology on *Car Cultures*, anthropologist Daniel Miller (2001:6) laments that both anthropology and sociology have neglected cars or automobiles to a "quite extraordinary degree, especially when compared to other examples of material culture such as food, clothing and the house." Since he made this assertion, while perhaps anthropologists and social scientists have not given automobiles and other vehicles the attention they deserve, there has been some progress in this domain, as is evidenced in the work of sociologist John Urry (2004, 2007, 2008) on automobility as a prominent form of mobility and in his book *After the Car* that he co-authored with Dennis Kingsley (Kingsley and Urry 2009). Unfortunately, Miller (2001:7) glosses over Freund and Martin's classic *The Ecology of the Automobile* (1993) in the introduction to his anthology, perhaps because it is more of a literary genre that tends to view the "car as a symbol of destruction," a stance of which he tends to be dismissive. While Freund and Martin highlight the social structural aspects of the automobile-industrial complex, they recognize that the automobile is not only a technological form for transportation but also a "cultural artefact in our personal experiences and our belief systems" which is viewed as an "inevitable and desirable feature of life" (Freund and Martin 1993:3).

Ultimately, automobility constitutes one domain, albeit probably the major one in at least developed societies, of what prompts people to be mobile, particularly in such a way that it requires some sort of technological device, be it an automobile, a motorcycle, bus, train, airplane, and even a bicycle.

Mokhtarian, Salomon, and Matan (2015:252) apply Abraham Maslow's classical hierarchy of needs to travel demand as is delineated below:

- Physiological needs: Travel for shopping for food or clothing or dining out
- Safety needs: Travel for work, health care, exercise, banking/investments, religious events, therapy, escape
- Social needs: Travel for social activities and recreation
- Esteem needs: Travel for status, independence, adventure, sense of conquest, escape
- Self-actualization needs: Travel for curiosity, restlessness, variety-seeking, aesthetic appreciation

Obviously, different needs can be specifically applied to private motor vehicles, but from an anthropological perspective they would be mediated by culture. Bearing this in mind, motor vehicle cultures consist of a wide array of subcultures and associated sub-societies, some of which I touch upon in this chapter. These include hot rodding groups, custom automobile groups, classic or vintage automobile groups, motorcycle clubs and gangs, and RV clubs and groups. Furthermore, numerous cultural tropes are associated with automobiles and other motor vehicles around the world. In this chapter, I examine nine of these, namely corporations as drivers of automobile culture; the automobile as a symbol of social status and self-worth; the automobile as a compensation for alienation in the workplace and social life; the automobile as a symbol of personal freedom, excitement, and adventure; the automobile as a symbol of gender and sexuality; the automobile as a symbol of national identity; the automobile as a symbol of ethnic identity; and the automobile as a living room and work space, and the glorification of private motor vehicles in novels, films, and songs. Needless, many of these tropes are not mutually exclusive but rather intersect with one another. In contrast to Miller (2001:8) who seeks to offer an "empathetic account of car culture consumption in particular cultural contexts," as is manifested by the various contributions to his anthology, my approach in this chapter and book tends to fall into a genre of writing that highlights the "consequences of the car." Furthermore, from a cultural perspective, automobiles are multi-faceted, contradictory, and paradoxical technological entities.

CORPORATIONS AS DRIVERS OF AUTOMOBILE CULTURE

The motor vehicle industry through advertising, lobbying politicians, and creating jobs contributes to automobile culture. As Barnet and Cavanaugh

(1994:262) so aptly note, "the car became a primary locus of recreation, a badge of affluence, a power fantasy on wheels, a gleaming sex symbol," all images that have been heavily promoted by the automobile industry through intensive advertising. In a similar vein, Redshaw observes:

> Manufacturers vie for position in the market place by appealing to the fantasy and pleasures of the car, and in the process are largely involved in framing articulations of cars. Car advertising frames the practice of driving in particular ways, playing a part in the shaping of expectations about cars and the experiences they are able to induce. (Redshaw 2007:125)

Alfred Sloan, the president and later the chairman of the board of General Motors from 1923 to 1956 sought to expand automobile sales with strategies such as "planned obsolescence" and purchasing a car on credit (Dauvergne 2008:40). He also offered "better quality" cars for aspirational consumers and introduced annual style changes to induce consumers to purchase the latest model and was essentially a pioneer in promoting aesthetically appealing automobiles. Not only upper- and upper-middle-class people were drawn to the aesthetics of automobility but also lower-middle-class and working-class people. Indeed in the 1920s GM began to offer mass-produced automobiles that resembled the more expensive models. In essence, whereas "Fordism" referred to the mass-production of automobiles, "Sloanism" referred to the advent of marketing techniques to sell mass-produced automobiles to a mass market, but one differentiated by its spending abilities. GM sold Cadillacs to high-income earners, Buicks to middle-income earners, and Chevrolets to working-class people. GM pursued a marketing strategy based on the premise "a car for every purse and purpose." Also, higher status people were more likely to purchase new models more frequently than lower status people.

> In one national study over half of all doctors, lawyers, and commercial travellers reported owning a car that was less than two years old. More than a third of physicians' cars were less than one year old. By contrast, half of all farmers, laborers, and other wage earners owned vehicles older than three years. . . . Cars "of extreme old," reported as nine years and older, were owned by the poorest Americans and used almost exclusively for economic activity. (Blanke 2007:44–45)

General Motors relied upon advertising, initially to provide an attractive alternative to the popular Model T Ford. By 1924, it "bought more magazine advertising than any other company in America, and it retained its position as America's largest advertiser for decades" (McCarthy 2007:89). Television provided GM and other automobile manufacturers with an outlet for proclaiming the wonders of their products, manifested for example, during the

1950s on the Dinah Shore show where she encouraged her viewers to "see the USA in your Chevrolet."

Slick advertisements in magazines, newspapers, and on television and billboards have been instrumental in selling automobiles. Psychologist John Cohen comments upon the quasi-religious aspects of automobile which seek to entice would-be customers: "A modern motor show carries all the trappings which in any other epoch would form the basis of a religious festival. It has colours, lights, priests (salesmen), kick off shoes, priestesses (fashion models), a ritual and a liturgy" (quoted in Marsh and Collett 1986:6).

Automobile manufacturers began to use advertising as early as the 1920s to market cars to women, portraying them as essential to domesticity and family life but also as a means for transcending the confines of domesticity and contesting gender stereotypes (Walsh 2012). In 2003 alone, the automobile industry spent around $10 billion on advertising in the United States (Koch 2012:118).

THE AUTOMOBILE AS A SYMBOL OF SELF AND SOCIAL STATUS

Daniel Miller maintains that people see and express themselves through their automobiles. The automobile is often viewed as an extension of the self, so much so that some scholars view the relationship between driver and car as a form of "hybridization" (Urry 2004). Wolf (1996:196) maintains that the "claim of the car society is that people in this uniform industrial society or in the new service industry society can express their individuality through their choice of car model." In a capitalist society, the private automobile constitutes the "individual means of transport par excellence," one that has become part of consumerist ideology, that along with home ownership, "instills an ideology of property so that the unpropertied masses should at least forget ideas of the genuine distribution of property" (Frank 1986:214). Thus, automobile ownership creates the illusion that all individuals are on an equal social footing.

This point aside, no one has personified the automobile as a symbol of social status and success more than Henry Ford who at one point was the richest person in the United States, possibly the world. The trappings of his success included the 56-room Ford Fair Lane mansion which included a bowling alley, an indoor swimming pool, a pipe organ, a library, not to speak of other amenities (Wik 1972:2–3). His estate was powered by electricity generated from a dam crossing a small nearby river. William K. Vanderbilt, the great grandson of Cornelius Vanderbilt and heir to the Vanderbilt fortune, in 1908 had a private 20-mile-long toll road constructed on Long Island which permitted him to drive his luxury automobiles at high speeds without incurring tickets for speeding and "provided a venue for the Vanderbilt Cup, one of the premier early American automobile races" (Wells 2013:41).

For others as well, the automobile "represents or symbolises hierarchical mobility," an indicator that one is moving up the ladder of social success (Redshaw 2007:124). As Schmidt-Relenberg (1986:124) observes, "In the public eye there is a hierarchy of car brands and types which corresponds to prestige scale of social positions." In their classic community study of Middletown (Muncie, Indiana) during the 1920s, Robert and Helen Lynd (1929:950) reported that the "make of one's car is rivalling the looks of one's place as an evidence of one's belonging' among members of the "business class." Further down the social ladder, many people in Middletown viewed acquisition of an automobile as a great personal achievement. Despite the Depression, the Lynds reported that by the mid-1930s, the car had become an essential object for the Middletown worker for whom "it gives the status which his job increasingly denies, and, more than any other facility to which he has access, it symbolizes living, having a good time, the things that keeps [*sic*] you working" (Lynd and Lynd 1937:245). In the United States, as middle-class and working-class people came to acquire automobiles in the early twentieth century, upper-class people turned to more expensive automobiles, particularly European brands such as Rolls-Royce, Mercedes, Hispano-Suiza, and Isotta-Fraschini (Gartman 1994:49). Overall, higher-income people tend to travel more than lower-income people, whether it be by automobile, train, or airplane. In the UK, Brand and Preston report:

> [P]eople in households with access to a car make more trips and travel further than those without access. . . . Thus income is a factor relating to the number of trips and distance travelled. In 2006, people in the highest income quintile did nearly 30% more trips than those in the lowest income quintile and travelled nearly three times further. . . . In particular, those in the highest income group did twice as many trips and travelled over three times further by car than those in the lowest income quintile group. (Brand and Preston 2010:1)

Numerous celebrities of various sorts have utilized automobiles as status symbols, marking their success. This has included religious figures, such as evangelist preacher Robert Schuller, the owner of a Rolls-Royce Phantom VI, who had initially established a drive-in church which eventually evolved into the Crystal Cathedral in Garden Grove in Orange County (southern California) (Marsh and Collett 1986:8). He actually presided over a ceremony in which Ruth Kramer Ziony, the editor of *Playgirl* "married" her Cadillac. In terms of car ownership, Schuller was up-staged by the Reverend Ike, an

African American Holiness preacher, who owned 19 Rolls-Royces and Bhagwan Shree Rajneesh, a guru who had to leave behind a fleet of 85 Rolls-Royces at his Oregon compound when he encountered difficulties with federal legal authorities. Elvis Presley perhaps more than any other celebrity used automobiles to validate his rise from his humble Mississippi roots to the King of rock and roll music. Widmer reports:

> As Elvis grew richer, automobiles became a type of personal currency for him and purchasing them bought him a peculiar form of economic self-expression. He bought all different types of cars: he bought many of them, and he bought them often . . . the self-made Sun King offered them freely to his attendants, and these munificent bequests served as informal salaries for his otherwise underpaid minions. (Widmer 2002:71)

In terms of political leaders, Arab oil tycoons enjoy purchasing custom-manufactured limos and sports car from dealers such as O'Gara Company in Los Angeles and Chameleon in London (Marsh and Collett 1986:24). Upper-class people and upper-middle class often collect historic, classic, or vintage vehicles, a marker of social status, consumption, entertainment, and even sociability given the existence of vintage associations, particularly in developed societies. If defined as motor vehicles over 30 years of age, statistics indicate that the UK has 805,588 historic vehicles in 2010, Germany 313,815 in 2013, Denmark 79,055 in 2012, Greece 402,932 in 2012, and Sweden 213,363 in 2013 (Araghi, Van Wee, and Kroesen 2017: 575). Historic vehicles tend to be much more polluting than modern vehicles, sometimes by a factor of five, but generally are driven for shorter distances.

In both rural and urban Zulu communities in South Africa, automobiles have in many instances replaced cattle as a status symbol, particularly for men but also for women. Jeske reports:

> For men in positions of power—government officials, pastors, teachers, business executives, or community elders—purchasing the best model of car possible not only displays individual status but also a social obligation to convey prosperity on behalf of a group. Not driving an appropriate quality of car dishonors the entire group represented and calls into question the leader's authority. (Jeske 2016:486)

Unlike traditional cattle, however, automobile ownership often proves to be a financial liability.

THE AUTOMOBILE AS A SYMBOL OF PERSONAL FREEDOM, CONVENIENCE, EXCITEMENT, AND ADVENTURE

Driving has been for some time synonymous with freedom in the United States but also elsewhere. As Foster (2003:67) observes, "[t]o teenagers and, indeed, for virtually all Americans automobility meant freedom." In its "Creed of the Open Road" published by the American Automobile Association (AAA) in 1932, it proclaimed that "love [for the freedom of the open road" was the primary impetus in the growth of automobile tourism (quoted in Blanke 2007:74). The automobile gave the rising middle classes a means for escaping the constraints of urban life, for example, in the form of auto touring and camping in mountainous and other scenic areas (Gartman 1994:35). The automobile allowed the upper and middle classes to escape the confinement of the cities and suburbs in order to visit state, provincial, and national parks, a luxury denied to the poor people and people of color. Keutcheyan observes:

> The nineteen-century emergence of "wilderness" was inextricably linked to the historically concomitant emergence of "Whiteness." The city was dark and dirty and that was where the dark and dirty individuals *par excellence* were to be found: Blacks, immigrants (Irish, Italians, Poles) and workers—these in fact often being the same people. (Keucheyan 2016:43)

On the other side of the world, a female Australian anthropologist colleague once told me that her car spells freedom for her as well, a sentiment that she shares with many other Australians in a land of wide-open spaces, at least outside of urban areas. Drivers view their automobiles as providing them with the freedom as well as convenience to travel by whichever route they choose as well as their destination of travel. Dennis and Urry (2009:39–40) argue: "The car's flexibility enables car drivers to get into their car and start it without permission or the expertise of others. It is ready and waiting to spring into life, so enabling people to travel at any time in any direction along the complex road systems that now link most houses, workplaces and leisure sites."

Based upon interviews with 15 Sydney drivers with a slightly higher than average household gross income, Kent (2015) found that one major reason that informants preferred driving to public transportation was that they could throw accessories needed for the day in the car rather than carrying them on public transportation or while cycling. Interviewees viewed their automobiles as providing comfortable and convenient space that shielded them "from others, and from the biophysical environment" (Kent 2015:737).

In Iceland, jeeps and four-wheel drive (4WD) vehicles provided inhabitants of this relatively remote island-nation beginning in the second half of the twentieth-century access to their country's most remote region, namely the highlands of the interior, a genre of travel that is "couched in the spirit of exploration, aiming to open up new terrains, drawing on an Enlightenment ethos" (Huijbens and Benediktsson 2007:151). Off course, Icelanders do not confine their jeep travel to the highlands in that jeeps are a common sight on city streets and rural roads.

An example of automobility that blends a sense of freedom, convenience, excitement, and adventure with sociability comes from Siberia where thousands of people from Novosibirsk pile into their cars from late spring to early autumn to drive eight hours on Fridays to the Altai Mountains, only to drive back home on Sunday (Broz and Habeck 2015). During their brief stay in the foothills of the massive Altai Mountains, for the most part they do not venture on hikes in the wilderness but merely take in the grand scenery of the area while consuming massive amounts of alcohol and oscillating between barbecue pits and saunas. Whereas during the Soviet era tourists made their way to the Altai region by regular or chartered bus trips, the latter often sponsored by their workplaces, the expansion of the culture of automobility in post-Soviet Russia has enabled people with some disposable income to participate in a culture of consumption in which the automobile represents the "promise of freedom, democratisation and modernisation" whereas public transportation "came to be associated with discomfort, poverty, and low social status" (Broz and Habeck 2015:565).

The above scenarios apply more to people in developed societies and the affluent in developing societies than they do for many poor people in developed societies and the vast majority of poor people in developing societies. Moreover, for even those who can afford to own automobiles, the "freedom of the road" might be constrained by congestion on the highway which may result from various factors, such as rush-hour traffic, inclement weather, or road accidents. Nevertheless, the belief that the automobile provides freedom is pervasive and consistent with neoliberal ideology which is part and parcel of late capitalism. As Boehm et al. (2006:13–14) observe, "despite the serious environmental, social and economic costs due to the 'success' of automobility, dominant political discourses call for cheaper fuel, less taxes, more roads and less governing of automobility." Conversely, in contrast to the neoliberal view of automobility as providing freedom, it entails an "entire physical, social and regulatory infrastructure to support movement along prescribed routes and modes" (Rajan 2006:118). Particularly in the segregated US South car ownership provided some African Americans an ersatz-freedom that entailed "winning a long-denied opportunity to shop on the same terms as other, more privileged citizens further up the wobbly ladder of racial hierarchy" (Gilroy 2001:86).

Along with the roller-coaster and the airplane, according to Duffy (2009:1) the automobile in terms of providing a new form of speed, constituted the most empowering and excruciating new experience for people everywhere in twentieth-century modernity." This is particularly manifested in the case of race cars and race car driving, which probably represents the epitome of excitement and freedom, both for the driver and the spectators. As Redshaw (2007:126) observes, "[t]he racetrack is the emblem of free expression—drivers can go as fast as their skills and vehicle will allow—while it is also controlled to minimise death and injury of drivers and spectators."

Motorcycles along with motorcycle gangs date back to the early 1920s and despite their association with disreputable characters, have been adopted by a wide array of people, ranging from hipsters in developed societies to those of limited financial resources in developing societies (Yates 1999). After World War II, motorcycles, which had been used as practical frontline vehicles during the war, made a comeback among civilians, including members of motorcycle gangs. While many motorcyclists view their personalized mode of transportation as perhaps the ultimate freedom-machine, the motorcycle in both the United States, Australia, and possibly at least in other developed societies, tends to be viewed by conservative individuals as a "threat to social order" (Packer 2008:8). Marquet and Miralles-Guasch (2016:44) report on the thrill of motorcycle driving in the congested Spanish city of Barcelona: "Driving in a dense and compact city such as Barcelona provides a dynamic experience for motorcycle rider, much in contrast with the static experience for car drivers. It is not just the objective shorter travel times, but also the pleasant experience of the free-flow movement between cars."

THE AUTOMOBILE AS A COMPENSATION FOR ALIENATION IN THE WORKPLACE AND SOCIAL LIFE

Paul Virilio (1991:65) maintains that the automobile provides people with an escape from reality, a "disappearance into a holiday where there's no tomorrow" and no destination necessarily. In a similar vein, motoring can be a viewed as a form of therapy that adjusts individuals to the "strains of modern life" (Ling 1990:4). Winfried Wolf states:

> The car society has shown itself to be an ideal appendage to both the bourgeois and the bureaucratic post-capitalist social formations. In both of these social formations, people are denied the most basic freedoms, for instance, the freedom to determine the products of human labour—what is produced, how, and

for what purpose. What they have are substitute freedoms which allow them to escape from the unfreedoms of day-to-day life. The car is one such substitute. (Wolf 1996:89–90)

Andrew Dawson (2017) conducted ethnographic research with a middle-aged Bosnian woman for whom driving serves as a therapeutic exercise which compensated from the trauma of her war-torn country. In her own words, Mira (pseudonym) stated: "Sometimes I feel like this is not my Bosnia anymore. . . . It is different when I am driving. . . . It's interesting, a discovery, manageable" (quoted in Dawson 2017:15).

Even in a generally affluent society such as Norway, there are individuals whose life chances are a bit marginal such as the women in their 20s in the small city of about 49,000 inhabitants in the south-eastern portion of the country. In her ethnographic study of these women, Pauline Garvey (2001) found that reckless driving and drinking formed a transgressive activity by which they expressed their alienation from mainstream society. For Kari, a 24-year-old woman who is unemployed and lives alone, her car "substitutes for absent relationships and she drives when she is lonely or bored—frequently at night while listening to music" (Garvey 2001:140).

THE AUTOMOBILE AND MOTORCYCLE AS SYMBOL OF GENDER AND SEXUALITY

Male engineers, mechanics, and drivers dominated early motor vehicle development and often viewed them as "boys' toys" designed to assert their masculinity. This process of particularly boys identifying with automobiles starts when play with toy cars and trucks, although sometimes girls may play with both as well, but clearly not as often. In some rare instances, boys decide to drive an automobile before they are legally eligible to do so. For example,

An eight-year old Ohio boy drove his sister, 4, to the local McDonald's after learning to drive by watching YouTube videos. They took their sleeping father's van for the 2.5 km journey. At the Maccas's drive-through window, the boy paid for the burgers with cash from his piggy bank. Police said no charges would be laid. (Odd spot 2017)

Along with other machines, male farmers added automobile repair work into their inventory of mechanical skills. As Kline and Pinch observe (1996: 779–780), "[t]he farm man's technical competence, rooted in his masculine identity, enabled him to reopen the black box of the car (by reinterpreting its function), jack up its rear wheels, and power all kinds of 'men's' work on the farm and, less frequently, the 'woman's' cream separator, water pump, or washing machine." Advertisements in hunting magazines tend to associate rugged masculinity with automobiles, guns, and dogs (Hirschman 2003).

Conversely, as Freund and Martin (1993:91) observe, the automobile is "unusual in its capacity to project both feminine and masculine imaginary, to carry erotic appeal for both women and men." In a similar vein, at least in Britain during the period of 1896–1939, O'Connell maintains that:

> It was commonly felt that men used cars for utilitarian and business reasons and they were therefore presumed to be attracted by cost, economy, ruggedness and reliability Less 'masculine' facets of the car were often associated with a growing feminine interest in motoring as a leisured pursuit, and women were adjudged to be attracted largely by styling, colour and comfort. (O'Connell 1998:64–65)

For the most part, automobiles project masculinity more than femininity by emphasizing power, speed, and aggressive driving. The Chevrolet Corvette with its sleek contours subliminally resembled the male penis. Marshall McLuhan (1964) refers to the automobile as a "mechanical bride" and maintains that it has "become an article of dress without which we feel uncertain, unclad, and incomplete in the urban compound."

Automobile shows have traditionally been events in which young, attractive, women, often attired in swim suits or evening gowns, are used to promote and sell automobiles. However, as I mentioned earlier, automobile companies targeted women in their advertisements as early as the 1920s. In the United States after World War II, suburban housewives came to increasingly rely upon automobiles in order to "shop at roadside strip developments or in burgeoning supermarkets and growing shopping malls that were difficult to access by foot" (Walsh 2012:216). Despite the stereotype of the inadequate "woman driver," during the 1950s more and more American women turned to driving, a pattern that developed in Europe and elsewhere somewhat later. As more and more women entered into the labor market, a second automobile came to be imperative for middle-class and working-class households. Thus today, both men and women, however, express their love for their automobiles.

Automobiles became a locus of romance and sexuality virtually from their earliest beginnings in that they permitted couples to have their interludes in new places, ranging from parks to dances and later drive-in theaters. Lewis reports:

> Early in the [twentieth] century many couples made love in cars because—in an era in which many young men and women lived with their parents—they had no better place to copulate. But others made love in cars because they found it exciting, sometimes dangerously so, and a change from familiar surroundings.

Lovers' lanes abounded in parks and off lesser streets and roadways in and around most communities. (Lewis 1983:132–133)

Automobile advertisements often convey sexual messages in a multiplicity of ways. In the 1980s, Peugeot placed an advertisement in *Playboy* magazine that asked its readers, "Do you leap into life first thing in the morning? ... Take a large boot size? Have a silken feel and a pleasing exterior? ... Does your mechanism respond to the faintest touch of a feminine hand?" (quoted in Marsh and Collett 1986:127). An Australian recreational vehicle company called Wicked Campers seeks to entice prospective renters with an advertisement on the rear of its vans stating "In every princess there is a little slut who wants to try it just once." In responding to a community campaign protesting the advertisement, the Queensland government announced that it would deregister the company's van if they did not comply with dictates of the Advertising Standards Bureau (Emanuel 2016:7). Council caravan parks in northern New South Wales have banned Wicked Camper vans and the Tasmanian government is considering taking similar action. Kamala Emanuel (2016:7) argues: The problem is that the Wicked Campers' slogan reinforces the Madonna/whore dichotomy women face as we negotiate the contradiction between being cast as sexually passive and respectable, on the one hand, or sexually experienced, assertive, adventurous and confident—and stigmatised as sluts—on the other." Motorcycle riding is perhaps the ultimate representation of heterosexual masculinity. Conversely, women have increasingly taken on motorcycle ownership and shifted away from their early role as a passenger sitting behind their man (Pinch and Reimer 2012:447).

THE AUTOMOBILE AS A SYMBOL OF NATIONAL AND ETHNIC IDENTITIES

Tim Edensor (2005:105) has highlighted automobiles as signifiers of national identity which constitute "familiar, iconic manufactured objects emerging out of historic systems of production and expertise" and the automobile industry as an "enduring signifier of national economic virility and modernity." The Rolls-Royce, even though it is now under German ownership, and the Mini encapsulate notions about Britishness. At a counter-hegemonic level, however, the British automobile industry has been "associated with militant trade union radicalism" (Edensor 2005:105). On the other side of the English Channel, Daimler Benz commanded a distinct status within German corporate culture during the Weimar and Nazi eras (Koshar 2005:123).

Although many societies, particularly in the developed countries, have viewed the automobile as a symbol of their national identity, this pattern is manifested in the United States more than any other society. French social theorist Jean Baudrillard (1988:54) asserts: "All you need to know about American society can be gleaned from an anthropology of driving behaviour." John Urry (2000:62) asserts that "American culture is in some ways inconceivable without the culture of the car." Christopher W. Wells (2013:289) refers to the United States as "car country" with a "monoculture, a landscape designed to maximize the benefits of car-based mobility." However, this monoculture has,

> for car-owning Americans, reduced once-formidable environmental constraints on easy travel to a distant memory, but one of the cruel ironies of Car Country is that other, very different environmental limits—including a destabilized climate and the depletion of crucial natural resource, oil—now pose the greatest threat to its future. The arena of Car Country's greatest successes has become its chief liability. (Wells 2013:295)

Ironically, while Sweden is a more culturally homogenous, politically and socially progressive, and secular society than the United States, Swedes tend to view their country as the most Americanized country in Europe, largely due to their consumption patterns, including their fondness for automobiles: During the 1950s, as O'Dell (2001:110) observes, "In the world of selected Swedish working- and middle-class dreams, the American car had a privileged place; it was the ideal which represented the beauty and potential of things to come." However, in the late 1950s and 1960s, many middle-class Swedes decided American cars were too big and turned to smaller European models, including Swedish ones. Conversely, certain young working-class men called *reggare* (greasers) adopted used American cars because they were inexpensive, but primarily because they represented an "alterity of their aesthetics" which "was partially a reaction against the dominant and normative Swedish preference for the practical and rational" (O'Dell 2001:114). *Reggare* even expressed their defiance of the Swedish middle-class during the 1970s when much of it protested the Vietnam war by "waving and prominently displaying American flags in their cars, while driving alongside and heckling marching anti-war demonstrators" (O'Dell 2001:114–115). In the case of another Scandinavian country, namely Norway, Pauline Garvey (2001:145) maintains that the "car appears as a core symbol of the state itself, in this case a Social Democratic state that claims to have brought affluence and progress while retaining its central role and progress while retaining its central role and its commitment to egalitarianism and welfare." In yet another Scandinavian country, the "assemblage of the jeep and its inhabitant is an active contributor to Icelandic politics of nature" (Huijbens and Benediktsson 2007:163).

Miller (2001) has written a fascinating account of the centrality of automobiles in one town in the developing world, namely Chaguanas in central Trinidad, the fasting growing urban center within Trinidad and Tobago. Most adult residents of Chaguanas managed to achieve affordability of a car, even if only a reupholstered older model, in the wake of the oil-boom of the late 1970s. In the process, the car came to dominate the Trinidadian identity:

> People are constantly recognized through their cars. . . . Street dialogue constantly asserts that men are attractive to women as much through the body of their cars as their own bodies and there are abundant metaphors based on car parts. (Miller 2001:286–287)

American society consists of various distinct automobile subcultures. Young white working-class American men particularly are "hot-rod" or "stock-car" enthusiasts, a role which allows them to exhibit their mechanical skills and talent for engaging in dangerous competitive driving. African Americans and Mexican Americans or Chicanos gravitated toward "low-riders," namely automobiles that hug the pavement. Working-class Asian Americans have embraced the "import street racer" (Seiler 2008:9). Historically, the possession of a Cadillac has been regarded by many African Americans as a symbol of success and their integration into mainstream American society (Packer 2008:190).

Marsh and Collett describe the sense of ethnic identity associated with belonging to a lowrider automobile club:

> Becoming a Lowrider and owning one of the street-hugging vehicles which have become the totem of the Chicanos is one of the main avenues through which a young Chicano can gain admiration and respect of his peers. As soon as he has amassed or borrowed the necessary finance, he begins the search for a second-hand automobile that can be customised according to the aesthetic canons of the lowrider culture. (Marsh and Collett 1986:104)

Anthropologist Brenda Bright (1998) conducted ethnographic research on low-riding in the Rio Grande Valley of northern New Mexico, particularly the Hispano communities of Espanola and Chimayo, about ten miles from each other. For adolescent males, lowriding constitutes an expression of personal and ethnic ethnicity associated with an emphasis on style and mechanical skills. Participation in lowriding has a gendered dimension: "While there are women who drive lowriders, men predominate and usually participate in some kind of organized lowrider group. In a few clubs, wives are allowing voting status, but for the most part, women do not form or join clubs. When they do join clubs, they tend to be active for a shorter period of time than men do" (Bright 1998:594). The existence of a famous religious shrine in

Chimayo has prompted the religious iconography on some lowrider automobiles depicting images such as the Sacred Heart of Jesus. Lowriders in the region "exemplify contemporary modes for the translocal productions of locality, demonstrating how local culture is produced in a global world" (Bright 1998:605).

The Anangu, an Indigenous people situated in the Western Desert of South Australia, treat their automobiles as "social bodies" which are used in a wide array of sociocultural activities, including hunting expeditions for emu and kangaroo, traveling to shops and work, to pick up mail and pay- and welfare checks, attend funerals, football matches, religious ceremonies, visiting family members and friends, and even avoiding social conflict within one's group by moving to another group (Young 2001). Men particularly, who generally are the "bosses" of cars, "may retreat alone to their cars for a smoke or to listen to music on tape or the local 5NPY radio" (Young 2001:42). The Anangu refer to discarded automobiles as "rubbish cars" and leave them on the roadside nose-to-nose with one another.

THE AUTOMOBILE AS A LIVING ROOM, WORK SPACE, AND ESCAPE FROM BOREDOM

Automobiles have for long been recognized as a home away from home so to speak. Duffy (2009:121) views the "car's space in twentieth-century culture often managed to eclipse and counter the home as site of sexuality, privacy, family gathering, and scene of generational changes." Some poor individuals and poor families in the United States even live in their motor vehicles not to speak of more affluent people, particularly retirees, who spend part of the year roaming the country in their recreational vehicles (RVs), often staying in RV parks where they find sociability with fellow RV enthusiasts. Bull (2001:199–200) observes: "Automobile users consistently refer to the car through the metaphor of 'home.' Yet a home in which they are preferably the sole occupant accompanied by the sounds of the radio or CD." The Waripiri, an Indigenous people, in the Northern Territory of Australia, often use their automobiles as a "mobile home and private bedroom: blankets and mattresses stored on seats, doubling as seat covers" (Stotz 2001:225).

Many modern automobiles, particularly luxury ones, have elaborate audio and even video systems, telephones, climate control for heating and cooling, and even ergonomically designed interiors. Dennis and Urry (2009:37) observe that the "car driver is increasingly surrounded by control systems that allow a simulation of the domestic environment, a home from home moving flexibly and riskily through strange and dangerous environments."

Many drivers will sing along with the music being played on their radio or CD. In many cases, drivers who live harried lives find listening to music in their motor vehicles therapeutic (Waitt, Harada, and Duffy 2017). The driver-car constitutes an "assembled social being that takes on the properties of a thing and a person and cannot exist without both" (Dant 2005:74). For some professional workers, their automobiles function as a moving office space in which they cut business deals on their phones and even keep files and other work-related supplies in the back seat and/or trunk (Laurier 2005). At least in developed societies, more and more drivers rely upon sometimes fragile GPS and GIS systems to instruct how to move from one location to another without consulting with a map. Architects have blended living space for an automobile into "carchitecture" by attaching garages and carports onto and into houses, and thus replacing the traditional porch in the process (Schiller et al. 2010:41).

THE GLORIFICATION OF PRIVATE MOTOR VEHICLES IN NOVELS, FILMS, AND SONGS

Automobiles and motorcycles have been an important theme in journalistic articles, novels, poetry, songs, television, and film. In terms of literature, John Steinbeck in *The Grapes of Wrath* refers to an "old and battle-scarred automobile long past its prime, as the new 'living principle,'" a means that promises to deliver his victims from the ravages of the Dust Bowl in the Great Plains to seek their future in the Promised Land in California, only to encounter new dilemmas (Dettelbach 1976:71). In Jack Kerouac's *On the Road*, Dean Moriarity finds his identity and sense of meaning in the process of traversing the United States back and forth in his car. In John Updike's famous *Rabbit* novels, Harry "Rabbit" Angstrom seeks escape from his humdrum life by constantly being on the run, often in his car. In literature automobiles often appear as symbols of material success, vehicles of power, dominance, control, and sexuality. This is the case in F. Scott Fitzgerald's *The Great Gatsby* (1925) in which Nick, Fitzgerald's protagonist, views Gatsby's Rolls-Royce as "emblematic of the new, money-made Eden, the status symbol par excellence of the fast-moving, free-wheeling society" but one which tragically kills an innocent bystander as its makes one of its hedonistic forays from the Gatsby's estate on Long Island to New York City (Dettelbach 1976:81). In Stephen King's novel *Christine* (1983), the protagonist Arnie falls in love with Christine, his car, a red Plymouth Fury which goes on a demonic rampage without her lover, including killing three boys who once abused her. Travelogues entailing motor vehicles include John Steinbeck's *Travels with Charley* and Che Guevara's *The Motorcycle Diaries*, which chronicles a journey undertaken with a close friend around South America that played a pivotal role in his political awakening, ultimately leading to his revolutionary activities in Cuba, Africa, and Bolivia. Volti asserts: "Infants at the end of the nineteenth century, but in their adolescence in 1920s they had become key components of an emerging consumer culture. . . . In the 1930s automobiles and movies were directly conjoined with the invention of the drive-in theatre, by which time cars had become featured players in movie scenes of every description.." (Volti 2004:292).

These include films, such as *Bonnie and Clyde* (1967), *Badlands* (1973), *American Graffiti* (1973), *Thelma and Louise* (1991), and many others. *Cars* (2006) portrays a world consisting of anthropomorphized automobiles and trucks. In contrast, *The Car* (1977) appeared at a time when, in American society, many progressive people had come to acknowledge the negative environmental and social impact of automobiles, portraying a driverless automobile that terrorizes the country as a symbol of the utopian future becoming converted into a dystopian future dominated by machines. Four *Mad Max* films, namely *Mad Max* (1979), *Mad Max 2* (1982), *Mad Max Beyond Thunderdome* (1985), and *Mad Max: The Fury* (2015) depict the themes of a dystopian world in an era of extreme oil shortage, the Australian outback, and Australian Aboriginality, the latter of which historically has been the antithesis of modernity but has been absorbed in large part by it (Gelder 1995).

In *The Wild One* (1953) Marlon Brando plays the role of a rebel without a cause in a film which also highlights the allure of motorcycle gangs. In contrast, *Easy Rider* starring Peter Fonda and Dennis Weaver features two hippie motorcyclists who sell a parcel of unidentified drugs to finance their journey across the United States whose small town residents take revenge upon the protagonists. In the *Little Fauss and Big Halsy* (1970), the protagonist, Little Fauss, bluntly says, "I just wanna race motorcycles and screw people." While road films are particularly common in the United States, they are quite common in various European countries, such as France and Germany (Archer 2017). Automobiles and other motor vehicles often provide a central motif in the films produced in Australia as well as in developing countries such as Iran.

Many of Elvis Presley's songs paid homage to American car culture. African American singer Chuck Berry also popularized auto-oriented lyrics during the late 1950s and 1960s. Mark Denning's "Teen Angel" (1960), Ray Peterson's "Tell Laura I Love Her" (1960), and Jan and Dean's "Dead Man's Curve" (1964) highlighted how automobile accidents brought young love to a tragic end. Various rock and roll groups, such as "The Chevelles," and "The Corvettes" adopted automobile models for their branding.

CONCLUSION

Over the course of the twentieth century and continuing into the twenty-first century private motor vehicles have become deeply embedded in modern life in both similar ways and different ways, depending on the culture and society. While their hegemony has been driven by motor vehicle companies, as we have seen in this chapter, private motor vehicles are multivalent, serving as symbols of selfhood and social status, personal freedom, convenience, excitement, adventure, gender, sexuality, and national and ethnic identity. They also provide compensation for alienation in the workplace and social life, as an ersatz living room and work space, and an escape from boredom. Automobility is an integral component of the social fabric of modern societies, particularly in developed societies but increasingly also in developing societies. It is a source of personal identity thus prompting some scholars for refer to the *autoself*, thus depicting drivers and automobiles as an articulated entity, a cyborg or human-machine hybrid (Randall 2017). Private motor vehicles evoke feelings, sentiments, fears, and anxieties, and serve as means to socially connecting with others. Private motor vehicles have been glorified in novels, films, songs, and in travel books throughout the world.

Motor vehicles also constitute a source of recreation, ranging from the classic Sunday drive into the country, professional people driving motorcycles on weekends, young men working on their hot rods, to people attending a race car event, be it the Grand Prix or a NASCAR rally. In North America and Australia, automobility fostered cultural icons such as fast food restaurants, motels, recreational vehicles, camping grounds within national and state or provincial parks and outside of them and what are termed "caravan parks" in Australia, and sprawling shopping centers.

Richard and Maurice McDonald set up a hamburger stand in San Bernardino, California, in 1940 that enabled them to serve drive-up customers (Foster 2003:92). Holiday Inns appeared in North America in the late 1950s but were followed by numerous other motel chains (Foster 2003:98). For those vacationers who wished to avoid spending relatively large amounts of money on staying in hotels and later motels, recreational vehicles began to provide accommodations beyond the simple or even elaborate tent. While initially some RVs were pulled behind automobiles, in time they evolved into large self-contained vehicles which consumed large amounts of gasoline. These self-contained RVs became particularly popular among retired couples who sometimes spend months on the road, traveling from RV park to RV park, and often accumulating new friends and acquaintances along the way. Schiller et al. report:

RVs often bring along off-road vehicles (ORVs) such as small motorcycles or snow machines, in order to explore and further pollute natural areas. The rapid increase of ORVs (especially in North America), whose emissions are generally less controlled than those of on-road vehicles, is a growing concern of environmental agencies. (Schiller et al. 2010:43)

Retirees in North America spend their sunset years traversing a huge continent, sometimes even selling their homes and living full-time in their RVs. In Australia, their counterparts are referred to as "grey nomads" and they may be found in caravan parks that dot the landscape.

Jessica Bruder (2017) chronicles the movements of another category of nomads which particularly have emerged in the United States in the wake of the Global Financial Crisis of 2008–2009. Due to dire economic circumstances, they have given up "traditional houses and apartments to live in what some call 'wheel estate'—vans, secondhand RVs, school buses, pickup campers, travel trailers, and plain old sedans" to roam the countryside shifting from seasonal to seasonal job along the way (Bruder 2017: xii). In the process, they form quasi-communities, meeting online, at a job, or camping off the electricity grid. Eschewing the label "homeless," Bruder argues that their real home is the open road, with their stopping-off spots being fast-food restaurants, all-night diners, shopping malls, Walmart parking lots, big box stores, automobile dealerships, megachurches, etc.

As the previous vignette indicates, the embeddedness of automobility and other forms of motor vehicle mobility is pervasive in modern life, not only in the United States, but in virtually all human societies around the world. As my brief discussion of cultural tropes associated with motor vehicles indicates, it is difficult to envision a world with far fewer motor vehicles. Yet, as subsequent chapters demonstrate, there are signs indicating that automobility as it presently exists, even though it continues to expand, is unsustainable in terms of space, particularly in urban areas, the natural environment, and health. Perhaps, in good time, many of the cultural tropes associated with motor vehicles will be replaced by cultural tropes associated with alternative forms of mobility, particularly walking, cycling, and train travel, which already exist, albeit in much more muted forms.

Chapter Three

The Political Ecological Impact of Motor Vehicles on Settlement Patterns

Over the course of the twentieth century and continuing into the twenty-first century, motor vehicles, particularly automobiles, have shaped the contours of cities, towns, and even rural areas as highways and streets of various sorts allowed them to make their way from place to place. The road became virtually synonymous with the car as other users, including in an earlier era, horse-drawn conveyances and up until today bicycles, were squeezed out (Agyeman 2013:111). Ironically, in the United States the League of American Wheelman, an organization of cyclists, lobbied for improving roads and along with farmers' groups and other interested individuals formed the backbone of the Good Roads Movement (Foster 2003:3). Albert Pope, the leading bicycle manufacturer in the United States, served as a leading Good Roads advocate. Furthermore, the movement was "backed financially by urban business interests, notably railroads and mail-order houses" (Ling 1990:39).

The increasing number of automobiles in cities in the early twentieth century altered the texture of urban street life particularly in Western societies. Indeed, pedestrians, parents, police, and downtown businesses often resisted their penetration, wishing to preserve street life as they had known it. Conversely, automobile interests "proposed in the 1920s that customary social constructions of streets were outdated and that only revolutionary change in perceptions of the street could ease congestion and prevent accidents" (Norton 2008:2). Historically, pedestrians could use streets as they wished, although sometimes they had to make way for horse-drawn carriages that conveyed aristocrats and other elites around towns and cities. The automobile, however, altered this mastery of the streets and imposed injunctions on walking in the middle of the road and jaywalking, In the United States, for instance, "Local Safety Councils advocated for safety in the new environment with lethal automobiles" (Garrison and Levinson 2014:198). Initially juries up until early 1920s sided with pedestrians in litigations over collisions with motorists, but the latter quickly gained the upper hand as automobiles defined the utilization of streets. Pedestrians were confined to sidewalks or footpaths and were theoretically forbidden to jaywalk, although this injunction to this day is frequently violated. By the 1930s, most pedestrians had come to accept the new social reality in which streets were defined as principally motor vehicle throughways, transforming the city into an automobile city. Even today, though some cities, such as Melbourne, have created bicycle lanes on roads, cyclists use these at their peril as cars, buses, and trucks generally whiz by them. Ironically, in congested situations, cyclists may whiz by motor vehicles. As Martin (2002:5) observes, "As an auto-centered transport system develops, it becomes the infrastructure of an auto social formation, which constitutes not only vehicles, roads, and drivers, but congestion, settlement sprawl, and so on, as well."

PUBLIC TRANSPORTATION, SUBURBANIZATION, AND LAND TRANSFORMATION

Automobiles replaced passenger trains to a large degree as both a mode of passenger transportation and long-haul trucks did the same to some degree as freight trains over the course of the twentieth century. In the United States, as Penna (2015:228) observes, "[r]ailroads had experienced almost 75 years of uninterrupted growth but, in 1920, they stopped growing." Largely due to the automobile, railroad passengers declined from some billion in 1916 to 700 million in 1930 and to 450 million in 1940 in the United States (Stover 1999:56). Whereas US railroad passenger-miles in 1929 were 34 billion, private automobile passenger-miles stood at 175 billion. Various US cities, particularly New York, Chicago, Washington, DC, and Baltimore, moved passenger railroad tracks underground in order make room for the growing number of motor vehicles on increasingly congested streets (Gordon 1996:297).

Urban and Suburban Areas

More than any other mode of transportation, over the course of the twentieth century the automobile became the "single most important influence on the configuration of urbanspaces" and contributed to creation of a "dispersed city," albeit this pattern varies considerably from city to city (Freund and Martin 1993:112). Architect Richard Rogers argues (1997:128): "[I]t is the car which has played the critical role in undermining the cohesive social structure of the city . . . they have eroded the quality of public spaces and have encouraged suburban sprawl . . . the car has made viable the whole concept of dividing everyday activities into compartments, segregating offices, shops and homes."

The automobile has contributed to urban and suburban sprawl. None other than Henry Ford was a proponent of this sprawl. He stated his anti-urbanist views as follows: "We shall solve the City by leaving the City. Get the people in the country, get them into communities where a man knows his neighbor, where there is a commonality of interest, where life is not artificial and cannot be made anything else" (Ford 1922:105).

This caricature of suburban life exemplifies the reality of automobile-dependent suburbs, whether they exist in the United States or elsewhere, including Australia.

At least in the North America and Australia, the suburbs preceded the automobile because they were initially served by trains and trolleys which allowed people to settle farther away from the congested city centers. As early as the 1920s, urban planning boards dominated by realtors, automobile dealers, and other special interests promoted road systems congruent with the culture of automobility. In the United States, as Kunstler (1993:90) observes, the "new low-density auto suburbs required expensive sewer and water lines to be laid *before* the new homes were sold—meaning that the carless urban working class had to pay for the new infrastructure that the car-owning middle class could enjoy." As more and more people adopted automobiles, particularly after World II, the automobile allowed people in the empty spaces and farmland between the train and trolley lines which developers were more than happy to transform into tract housing, creating suburbs such as Levittown on Long Island. Like Henry Ford, the renowned American architect Frank Lloyd Wright was also a proponent of suburbanization. In his utopian vision of urban life called Usonia portrayed in *The Living City* (1958), he argued that "[c]ars (and personal helicopters) would enable Usonians to place more comfortable distances between neighbors, freeing themselves from the typical big-city streetscape" (Owen 2009:108).

Even in Western Europe which has had better public transit systems than North America and Australia, the automobile has had a profound impact on urban and suburban settlement patterns. As Freudendal-Pedersen observes,

> Today, cities are organized according to the architecture of automobility. Contemporary mobility, particularly automobility, takes up a huge amount of space in the city and creates congestion and insecurity. Today 25 percent of the land in London is a car-only environment, a figure similar to that of several Nordic countries. It seems that when a car is acquired, most trips are facilitated by automobility. (Freudendal-Pedersen 2009:4)

Around the world vast tracts of land, much of it farmland, are being transformed annually for motor vehicle use in the form of roads, highways, and parking lots. A powerful lobby consisting of the automobile industry, petroleum companies, and trucking companies historically has posed a formidable barrier to the development of efficient public transportation systems in many countries, particularly in US urban areas. In the case of the United States, Avila (2014:26) includes insurance companies, producers of rubber, glass, and steel, suburban retailers, housing developers, and advocacy groups such the Automotive Safety Council, the National Automobile Association, the American Association of Highway Improvement, and the Urban Land Institute as being actors within the automobile lobby, or the "Road Gang."

A pattern of motor vehicle-related industries manipulating the market in such a way to destroy tram and suburban train systems began as early as the 1920s when a General Motors subsidiary bought out streetcar systems in Springfield in Ohio and Kalamazoo and Saginaw in Michigan and pressured reconstituted transit systems to purchase only GM and Mack buses, Firestone tires, and fuels and lubricants from Standard Oil of California (Garrison and Levinson 2006:127). A consortium called National City Lines, consisting of General Motors, Standard Oil of California, and the Firestone Tire and Rubber Company, spent $9 million by 1950 to obtain control of trolley or tram companies in 16 states and converted them to less efficient GM-manufactured buses. The companies were sold to operators who signed contracts specifying that they would purchase GM equipment. National City Lines in the 1940s began buying up and scrapping parts of Pacific Electric, the world's largest interurban electric rail system, which by 1945 served 110 million passengers in 56 smog-free Southern California cities. Eleven hundred miles of Pacific Electric's track were torn up, and the system went out of service in 1961, as Southern California commuters came to rely heavily on freeways (Flink 1973:220). Unfortunately, Henry Huntington, the owner of Pacific Electric, had used his interurban trolley company largely as a scheme for promoting his real estate endeavors rather than providing a public service, thus often alienating customers in various ways, including his failure to provide lines that connected suburbs to each other as opposed to strictly city centers.

A similar process in which a consortium of road interests colluded to destroy efficient trolley or tram systems occurred throughout the United States and Australia (Goddard 1994; Davison 2004). General Motors and its allies destroyed more than 100 trolley systems in numerous North American cities, a process in which Ford and Chrysler also played a role in the eroding the quality of public space and contributing to suburban sprawl. General Motors, Mack Manufacturing (trucks), Standard Oil, Philips Petroleum, Firestone

Tire & Rubber, and Greyhound Lines shared information to eliminate public transportation, particularly in the form of trolleys or trams, and essentially violated anti-trust laws between 1927 and 1955 (Dennis and Urry 2009:35). In contrast to 1929 when some 14.4 million people traveled by trolley, due to the shift to cars this number had by 1940 nearly halved (Kay 1997:213). In 1935 most of the trolleys in Manhattan had been replaced by buses. The US Supreme Court convicted National City Lines, Pacific City Lines, Firestone, Philips Petroleum, Mack Trucks, Standard Oil, and some other companies along with two individuals—namely, Roy Fitzgerald and H. C. Grossman—of having violated anti-trust laws, but fined each of the corporations a mere $5,000 and the two individuals $1 (Goddard 1994:135). Whereas large US cities such as New York, Boston, Philadelphia, Washington, DC, Chicago, and the San Francisco Bay Area, developed and still operate extensive and relatively efficient public transportation systems, Detroit—the capital of the motor vehicle industry—has relied since the demise of its trolley system primarily on buses for its public transit system. By 1955 General Motors had achieved its goal of motorizing US cities, a strategy that crippled most trolley, bus, and trolley bus systems (Goddard 1994:135). Various scholars, such as Dunn (1998), have argued that GM and its allies did not destroy trolleys, contending that they only took advantage of economic trends already set in place by the advent of motorization and suburbanization. Such arguments, however, unfortunately downplay the role of corporate influence on public policies and corporations historically have colluded to serve to their own mutual interests.

In the 1950s, with the assistance of the Eisenhower administration, the development of an interstate highway system resulted in enormous profits for corporations and benefits for supportive politicians, while hindering the development of public transportation, and thereby forcing the public to purchase and use cars. Indeed, Lewis Mumford (1963) argued that the federally funded highway programs of the 1950s contributed to the creation of a "one-dimensional transportation system." John F. Kennedy, a Democrat, followed the Eisenhower administration by urging completion of the interstate system (Jackson 2003:212). The Association of American Railroads mounted a huge public relations campaign in 1967 whereby it "ceded passenger traffic to roadways and pressed railway's advantage in carrying freight" (Jackson 2003:215). However, the railroads found themselves in stiff competition with the trucking industry. According to Crawford (2002:88), "[t]he Interstates gave truckers a subsidized route network that allowed them to compete successfully with railroads despite the labor and energy inefficiency of trucking. It also gave real estate developers the high-speed arteries that made large-scale suburban sprawl possible." Geographer David Harvey (1990) maintains that highway systems have played a crucial role in the urban growth of late capitalism both in the United States and elsewhere.

While interstate highways obviously made long-distance driving much easier, in urban areas they contributed to "white flight" by separating European Americans from African Americans and Mexican Americans (Lewis 1997:x). They also served to isolate small towns not close by, often resulting in the demise of their commercial life as gasoline stations, restaurants (including franchises of Howard Johnsons, McDonald's, etc.), and motels increasingly situated themselves near freeway on-ramps. Housing developments and shopping centers in time also became situated near the interstates (Lewis 1997:282). The interstates also facilitated a tremendous expansion of shipping by trucks, which reportedly increased 257 percent between 1955 and 1990 (Lewis 1997:286). Industries shifted their operations from multi-story buildings in CBDs to single-story buildings on wide roads in suburban areas, thus facilitating easy access to the interstates.

A powerful lobby consisting of the automobile industry, the American Automobile Association, petroleum companies, trucking companies, and tire companies continues to pose a barrier to the development of efficient public transportation in much of the United States. Whereas heavy trucks contribute more than 95 percent of the highway deterioration in the United States, trucking firms pay only 29 percent of the country's highway bill (Freund and Martin 1993:2).

Numerous US cities, such as Los Angeles, Houston, Detroit, Atlanta, Dallas-Fort Worth, Phoenix, as well as cities in other parts of the world, such as Perth and Canberra in Australia, and Bangkok in Thailand, have developed into what Newman and Kenworthy (1999:31–33) aptly term "automobile cities." In contrast to European societies, as Flink (1988:374) observes, the "federal, state, and local governments in the United States have consistently provided massive funds for building the world's best highway infrastructure, to the virtual exclusion of aid for the rail infrastructure." Car dependency is particularly pronounced in North American and Australian cities. Table 3.1 depicts the various modes of transportation that people use to travel in selected cities in these two regions. While North American and Australian cities vary widely in terms of the quality of their public transport systems, except in New York, the vast majority of people in them opt to travel to work by car, even when they have ready access to public transport facilities.

Table 3.1. Methods of Travel to Work in Selected North American and Australian Cities

City	Car (%)	Public Transport (%)	Walking (%)	Cycling (%)	Other (%)
Los Angeles	91.1	4.7	2.7	0.6	1.1
Toronto	71.1	22.2	4.8	1.0	0.9
New York	20.5	67.6	24.8	0.3	1.6
Vancouver	74.4	16.5	6.3	1.7	1.1
Melbourne	79.3	13.9	3.6	1.3	1.9
Phoenix	93.4	1.9	2.1	0.9	1.4
Canberra	82.0	7.9	4.9	2.5	2.7

Source: Adapted from Mees (2010:60-61).

In terms of passenger miles in 2007, Garrison and Levinson (2014:329) provide the following breakdown which again illustrates the automobile dependency in the United States:

- Passenger cars and motorcycles, 49.2%
- Trucks, 35.83%
- Air carrier, 11.30%
- Inner-city/Amtrak, 0.11%
- Other, 3.35%
  - Buses, 2.75%
  - Rail transit, 0.34%
  - Commuter rail, 0.21%
  - Other transit, 0.06%

American cities tend to be the most automobile dependent of those depicted above, followed by Australian cities, then Canadian cities, European cities, and finally two modern Asian cities, both of which have excellent public transit services. Bear in mind that SUVs and pickup trucks that are used for general driving rather than for work-related purposes fall under the rubric of "trucks." In the case of Australasian cities, Newman, Kenworthy, and Bachels (2001:56) provide the following breakdown in terms of automobile dependency:

- Extreme auto dependency
  - Perth, Adelaide, Canberra
- High auto dependency
  - Christchurch, Brisbane, Auckland, Melbourne

- Moderate auto dependency
  - Sydney, Wellington

Even when people do not use their own private automobiles, they may rely on automobiles in the form of taxicabs when they are away from home on business or holiday. In the late 1990s in the United States, the "most automobile-like public transit mode, taxicabs . . . [carried] more passengers than all other kinds of public transit" in total (Dunn 1999:15).

Modern cities have evolved following, in large part, the dictates of capital with its need for manufacturing, financial, commercial, distribution, and communication centers, as well as state bureaucracies. In cities where significant public transportation infrastructure exists but has not been developed and upgraded sufficiently to discourage car use, the growing proliferation of cars also reduces the efficiency of road-based forms of public transportation. In Melbourne, for example, buses and trams are slowed greatly by congestion caused by cars. Fortunately, a few Melbourne trams operate as light-rail conveyances in some motor vehicle-free stretches between the city or central business district and various suburbs.

The automobile has contributed immensely to suburbanization and settlement patterns of cities around the world. Parissien reports:

> Between 1945 and 1954, nine million Americans moved to the suburbs, and by 1976 more Americans lived in suburbs than in downtown or rural areas, seeking the space, safety, autonomy, greenery and cleanliness that suburban life promised. The first planned out-of-town shopping centre opened in Raleigh, North Carolina, in 1949: the first enclosed, climate-controlled shopping mall appeared in Minneapolis in 1956; and by 1980 there were over twenty thousand major suburban shopping centres across the US. (Parissien 2013:189)

Lester Brown (2001) calculated that the United States has 3.9 million miles of roads, "enough to circle the earth at the equator 157 times . . . and devoted, an expanse approaching the size of the 51.9 million acres that US farmers planted in wheat last year."

Transportation planners particularly in North America and Australia have designed roadways that seek to keep traffic moving through an elaborate system of freeways and other high-speed highways, which more recently came to include ring-roads around the city center. European and even developing cities have to some extent followed suit. The World Bank has facilitated this process in various Chinese and Indian cities as well as other developing cities (Newman and Kenworthy 2015:142).

In North America, Australia, and to some extent Europe the automobile has contributed to the demise of traditional working-class enclaves in the inner cities which contributed to unionism and radical politics and the emergence of a homeowner culture that dispersed the working class into bedroom suburbs where they participated in the culture of consumption (Koch 2012:73). It has facilitated access to shopping malls where seemingly an array of consumer goods could be obtained in stored into larger and larger dwelling units. During the Global Financial Crisis of 2008–2009, the rising price of oil transformed many US suburbs into "'ghostburbs,' full of foreclosures, for-sale signs, and empty housing" (Urry 2016:117).

Hickman and Banister (2014: 179) delineate a spectrum of city types between the "car city" and the 'transit city" depicted below: noting that most of them actually constitute hybrid forms:

- Motorcycle cities – Hanoi, 1980s
- Traffic saturated motorcycle cities
  - E.g.: Ho Chi Minh City, Hanoi in the 1980s
- Motor car cities
  - E.g.: Los Angeles, Houston, Dubai
- Car cities in decline
  - E.g., Detroit
- Bus/paratransit cities
  - E.g., Seoul, Manila in 1970s
- BRT [bus rapid transit] cities
  - E.g., Curitiba, Bogota, Jakarta, Jinan, Lagos
- Transit cities
  - E.g., Hong Kong, Shanghai, Singapore, Tokyo. London, Paris, Zurich
- Modern cycling cities
  - E.g., Amsterdam, Copenhagen, Oxford

Newman and Kenworthy (2006:35) maintain that based upon long-term data from cities around the globe there appears to be a "fundamental threshold of urban intensity (residents and jobs) of around 35 per hectare [one hectare equals 2.47 acres] where automobile dependence is significantly reduced.

Freund and Martin (2007:37) refer to the high level of individualized and intensified movement by car as *hyperautomobility*, which includes increased solo driving, more trips, greater trip distances, and suburban sprawl in communities characterized by "low-density, fragmented sites (i.e., single-family housing, shopping malls, and corporate campuses) that favour automobile travel over other transport modes." Hyperautomobility contributes to various social and public health problems, such as social isolation, marginization of public transportation, and decline of cycling and walking which contribute in turn to obesity and hazardous road conditions for pedestrians.

The amount of land devoted to automobiles in cities around the world is incredible. In the CBDs of US cities, approximately half of land surface space is devoted to streets and parking lots, two-thirds in the case of Los Angeles's CBD (Wright 1992:35). The figure for London's CBD is 22 percent and for New York's 24 percent.

Rural Areas

Small towns and rural areas are generally even more car dependent than urban or suburban areas. In developed countries, car ownership tends to be higher in rural areas than in urban areas, in large part due to inadequate public transportation. Indeed, railway and bus services to rural areas have significantly declined with the advent of automobiles not only in the United States but also the UK and various European countries. Bus ridership in the UK in terms of passenger journeys more than halved between 1955 and the late 1990s (Hoyle and Knowles 1998:189). In most parts of the UK, Sunday bus services have disappeared. About 95 percent of rural residents in the United States own cars, "with two-thirds of rural households having two or more cars (Hoyle and Knowles 1998:199). Most rural towns in the United States went directly from horse-drawn carriages directly to automobiles. Whereas many US rural towns were served by trains, albeit often on a sporadic basis, this generally is not the case today. I have been struck by the hegemony of automobile dependency in small rural towns in the United States and Australia, even when points of destination can be easily reached by walking and particularly cycling. During my teaching stint in 1972–1973, at Kearney State College in south-central Nebraska, an institution of some 6,000 students, some of the students who lived a few or several blocks from the campus, opted to drive there and then drive around looking for a parking space. Very likely for them owning a car was more a mark of social status than convenience, but this is an issue that warrants further study. Even some of my male Nigerian students who did not have much money aspired to own a car, a symbol of American modernity. While conducting research on a health care program in rural Utah towns in summer 1975, I resided for a month in a basement apartment with my former wife and our baby son in Monticello, a town of some 1,400 residents at the base of the Blue Mountains in southeast Utah. Our landlady worked at a grocery store about a five-minute walk from her house. Instead of walking to her workplace, she drove there in the morning, drove home for lunch, drove back to work, and then drove home at the end of the work day. Bear in mind, there are many urbanites in American and Australian cities who drive to a nearby store rather than walk there as well. Given that many European villages predate the advent of the automobile, they were designed in large part as pedestrian communities and often to some degree remain so. Unfortunately, many North American and Australian small towns have a long thoroughfare that connects one end of town to the other, with businesses often strung out along the thoroughfare. Fortunately, while Australian small country towns of up to a few thousand residents, if they have a passenger railway station, often near the CBD, they can be quite walkable. communities. Obviously, people who live relatively long distances from rural towns on farms or ranches almost always are motor vehicle dependent.

Land and Motor Vehicles

Motor vehicles require an enormous amount of land for highways, parking spaces on streets, and parking lots and garages. Kay (1997:64) maintains that the United States has become a "hard-topped world": "From 30 to 50 percent of urban America is given over to the car, two-thirds in Los Angeles. In Houston the figure for the amount of asphalt is 30 car spaces per resident." In the case of Britain, Martin reports:

> In rural settlement, more vehicular traffic expropriates more landspace: Roads are widened and bypasses are built. Yet, congestion still spills over to secondary and tertiary roads. Country lands in the English home countries have become increasingly difficult to navigate for their customary users—walkers, cyclists, equestrians, farm vehicles, and livestock. (Martin 2002:10)

By the late 1990s, the US Forest Service had developed some 370,000 miles of road on its 300,000 square miles of jurisdiction (Kay 1997:83). In Eastern European countries, which once relied heavily upon trolleys as a form of public transport, the wave of consumerism that followed the collapse of the Soviet bloc resulted in a massive increase in the number of automobiles as well in traffic congestion. East European countries embarked upon massive programs of highway construction and abandoned railway lines as more and more people turned to cars for transport.

China has been feverishly building freeways in both the countryside and cities with over 21,000 miles (32,000 kilometers) traversing the country in 2004 and an expected doubling of this network by 2020 (Sperling and Gordon 2010:2009–2010). Freeways and roads in the countryside still tend to be relatively lightly traveled, with a heavy concentration of trucks and buses, but those in the cities, particularly large cities such as Beijing and Shanghai, have become congested with all sorts of motor vehicles. In contrast, the Chinese government has been slow in constructing commuter train and subway systems to accommodate rapid urbanization. Conversely, the Chinese state has begun to recognize the problems associated with favoring motorized vehicles over bicycles and pedestrians in large cities. Shanghai now encourages bicycle use and prioritizes public transportation, "so much so that car ownership is limited by a system of auctioned entitlements to purchase cars, similar to Singapore's system, that keeps Shanghai's car ownership rate far below Beijing's (37 vs. 108 passenger vehicles per 1,000 persons)" (Gilbert and Perl 2010:75). Chinese cars tend to be smaller than both North American and European ones.

In addition to China, the paradigm of automobility has become hegemonic in numerous other developing societies around the world. Pakistan, for example, is constructing more and more roads to accommodate a "tiny minority of aspiring motorists" (Low and O'Connor 2013:3).

Congestion

Cities around the world vary considerably in their utilization of cars for transportation. In general, as more and more people in an urban area begin to use motor vehicles coupled often with a growing urban population, traffic congestion inevitably results. Newman and Kenworthy describe the logic behind this development:

> Once cities are locked into a primarily road-based system, a momentum develops that is hard to stop. The typical response to the failure of freeways to cope with traffic congestion is to suggest that still more roads are urgently needed. The new roads are then justified again on technical grounds in terms of time, fuel, and other perceived savings to the community from eliminating the congestion. This sets in motion a vicious circle or self-fulfilling prophecy of congestion, road building, sprawl, more congestion, and more road building. (Newman and Kenworthy 2015:143)

Various indices have been used to measure traffic congestion. The RCI was developed at the Texas Transportation Index. The RCI takes the following form:

$$RCI = \frac{(FwyVMT/Ln\text{-}Mi) \times FwyVMT + (ArtVMT/Ln\text{-}Mi) \times ArtMT}{(13,000 \times FwyVMT0 + (5,000 \times ArtVMT)}$$

where FwyVMT/Ln-Mi is the freeway daily vehicle miles per lane travelled, FwyVMT is the freeway daily vehicle miles travelled, ArtVMT/Ln-Mi is the principal daily vehicle miles travelled, 13,000 is the capacity per lane on freeways, and 5,000 is the capacity per lane on principal arterials. (Black 2010:66)

Table 3.2 below depicts how this has occurred in selected US cities between 1982 and 2007.

Table 3.2. Roadway Congestion Index for Selected US Cities

City	Roadway Congestion Index		
	1982	1994	2007
Atlanta	0.83	1.18	1.31
Chicago	0.81	1.03	1.18
Cleveland	0.73	0.89	0.89
Detroit	0.91	1.12	1.23
Los Angeles	1.21	1.49	1.58
New York	0.73	0.93	1.15
Pittsburgh	0.67	0.73	0.78
Washington, DC	0.83	1.21	1.34

Source: Adapted from Black (2010:67).

While the sample of cities depicted in table 3.2 represents only a small portion of large metropolitan areas in the United States, the RCI increased in all of them between 1982 and 2007, although the increase was relatively small in two former steel-producing cities, namely Cleveland and Pittsburgh. The average American spends 47 hours each year in congestion, costing him or her over $1,000 a year in gasoline and wear and tear on his or her automobile (Balaker and Staley 2006: xiii). Congestion also interferes with quality time with one's family and significant others and meeting a romantic partner. One survey revealed that thousands of Atlanta-area Match.com (an internet dating site) subscribers would not date anyone who lived more than 10 miles away.

Based upon research conducted at the Texas Transportation Institute, Staley and Moore (2009:16–17) report that while in 1982 Los Angeles was the only US urban area facing "severe" congestion, this characterization applied to 32 US urban areas in 2005 and potentially could apply to 52 US urban areas by 2030 if drastic measures would not be implemented. Rush-hour motor vehicle speeds have been reported to be seven miles per hour in London, 12 miles per hour in Tokyo, 17 miles per hour in Paris, and 33 miles per hour in Southern California (Freund and Martin 1993:2). For instance, car or motorcycle utilization runs from 16 percent age-eligible residents in Hong Kong, to 39 percent in Berlin, to 51 percent in Madrid, to 76 percent in Melbourne, to 88 percent in Chicago (Gilbert and Perl 2010). In China, car ownership varies widely from area to area: "In Beijing, there were 108 cars per 1,000 residents in 2004, 72 percent were privately owned. In Gansu, Guizhou, and Jiangxi provinces, there were fewer than 5 per 1,000, and only 42 percent of these cars were privately owned" (Gilbert and Perl 2010:74).

Even in Singapore, which reportedly has an excellent public transportation system, many affluent people are being seduced by automobility. In this densely populated and compact city-state of only 274 square miles,

> Mileage per vehicle is very high: the average car in Singapore travelled 21,100 km in 2006, double the figure for 1980 and virtually identical to the 21,317 km reported for Los Angeles. . . . The number of Singapore residents rose by 6 per cent in the five years to 2007, but the number of cars jumped 19 per cent, traffic entering Singapore's CBD grew by 14 per cent and CBD traffic speeds fell by up to 30 percent. (Mees 2010:45)

Not only in Singapore but throughout much of Southeast Asia, South Asia, and East Asia, cities are increasingly characterized by traffic congestion in which motor vehicles make their way across large cities such as Bangkok, Jakarta, Beijing, Shanghai, Mexico City, and Sao Paulo in a long series of starts and stops. Lower-income cities in developing countries are undergoing a chaotic evolution from their former status as walking cities to public transport and still later to automobile cities. Traffic congestion in large cities in developing countries tends to be worse than it is in developed societies, in part due to poor public transportation systems in the former. In contrast to developed societies where cars constitute the primary private mode of transportation, towns and cities in some developing countries teem with motor scooters, motorcycles, and mopeds because they generally are considerably less expensive than cars (Metz 2008).

Arndt et al. conducted an interesting study of the transport modes in five megacities in developing countries and found a variegated "modal split" which is depicted in table 3.3 below:

Table 3.3. Transport Modal Split (%) in Five Selected Megacities in Developing Countries

Megacity	Pedestrians	Bike	Public Transport	Motorcycles	Car	Other
Tehran (Iran)	4	2	55	7	32	—
Hyderbad (India)	30	3	28	31	2	6
Ho Chi Minh City (Vietnam)	3	1	6	81	9	—
Gauteng Province (South Africa)	35	2	27	1	35	—
Hefei (China)	30	22	19	5	10	14

Source: Adapted from Arndt et al. (2014:19).

Of all the cities in the study, Tehran (population of nearly 13.8 million) has the best public transportation system, one which is highly utilized (Arndt and Doege 2014). Hyderbad, with a population of some 6.8 million, has the highest number of Special Economic Zones of all Indian cities and has therefore been experiencing tremendous economic growth, which has spurred an exponential increase in private motorized vehicles, particularly motorcycles. Asto Schaefer and Jain (2014:42) observe, "this is caused by the massive urban sprawl of Hyderabad—which has resulted in increased trip lengths for most urban residents—deterioration of PT [public transport] coverage, and simultaneously, the decrease in pedestrian and bicycle traffic."

Of all the megacities in the study, Ho Chi Minh City (official population of 8 million in 2011) has by far the highest motorcycle utilization rate. Emberger reports:

> Private transport is based on motor scooters, because motor scooters are flexible, fast, and affordable. Parking is permitted everywhere in the city and is relatively inexpensive. Since the overall speed of motor scooters is between 15 and 20 km/h, a rather unusual kind of traffic flow results, giving newcomers the feeling that the whole road network is in steady motion with hardly any rules. Nowadays, the increasing rate of private ownership seems to have disturbed the flow of scooters, thus making traffic rules, and their enforcement even more necessary to ensure movement in the chaotic mix of scooter and car traffic in HCMC. (Emberger 2014:26–27)

HCMC's public grossly underdeveloped public transport system is based upon buses, most of which are old.

Gauteng province constitutes a sprawling suburbanized area in South Africa with a population of some 11 million people, roughly a 1/5 of the country's population. Tomaschek and Fahl report:

> Gauteng's transport sector is highly dependent on motorised individual road and freight road transport. Historically, the public transport system was not well supported by the government, and apartheid city planning caused urban sprawl with people living far away from their workplace. Public transport was mainly developed for long-distance commuter rail, which was intended to bring the poor population living outside the city centres to their workplaces. Only minimal commuter bus routes were provided to the city centres. An unregulated taxi system has filled the gap by the lack of an efficient public transport system. (Tomaschek and Faul 2014:78)

While Lah et al. (2014) were part of the same study as were Arndt et al. (2014), the statistics that the former provide in their chapter differ somewhat from those reported by the latter, as indicated in table 3.4 above for the Hebei province (population about 5.3 million), which is on the verge of evolving into a full-blown megacity. At any rate, Lah et al. (2014:98) report that in 2010 the primary transport in Hebei consisted of walking 30.34 percent, push bicycling 21.98 percent, electric bicycling 14.26 percent, taxi 2.36 percent, public transit 11.29 percent, and automobile 11.29 percent. In 2010, Hebei's vehicle fleet consisted of some 200,000 automobiles, 8,855 taxicabs, 2,896 buses, some 10,000 E-bikes, 46,190 converted motorbikes, and 17,600 HGVs (Lah et al. 2014:99).

Low-income cities exhibit what Kenworthy (2011:107–108) dubs the "Bangkok syndrome" which includes the following dimensions:

- A tendency to rely on motorized transport rather than public transportation
- A tendency to resort to developing roads rather than public transportation in new urban developments
- A tendency for public transportation to rely on buses which become bogged down in congestion rather than rail systems
- "Congestion is being exacerbated by the auto-oriented form of a lot of new urban development on the fringe"
- "More significant roles for motorcycles and taxis in the modal mix, at least partly due to inferior formal public transport systems"
- High pollution created by urban transportation due to heavy reliance on private transport modes, especially high reliance on motorcycles, poor vehicle maintenance, and emissions regulations
- High transport deaths per capita resulting from a "combination of poor-quality roads, poor driver education, inferior vehicle safety standards and often lax law enforcement"

While only a few low-income cities have trams, like buses, they move very slowly due to heavy congestion. Each motorist in Bangkok, for example, reportedly spends an average equivalent of 44 days per annum in traffic jams (Banister 2005). In short, urban areas around the world find themselves to a greater or lesser degree figuratively inundated by private motor vehicles of one sort or another, in terms of air pollution, noise, greenhouse gas emissions, and congestion that severely counteracts many of the redeeming aspects of urban life, such as cultural diversity and cosmopolitanism.

It is important to note that there are cities in developing countries that have highly efficient public transit systems and do not necessarily suffer from the Bangkok syndrome. Buenos Aires in Argentina, Caracas in Venezuela, Hong Kong, Mexico City, Santiago in Chile, and Singapore have extensive Metro systems; Kuala Lumpur in Malaysia, Manila in the Philippines, and Bogota and Medillin in Colombia have light rail systems; and Curitiba and Sao Paulo in Brazil and Quito in Ecuador have efficient bus rapid transit (BRT) systems (Banister 2005:202). Nevertheless, while not necessarily being automobile-dependent cities per se, many cities in developing countries, such as Taipei in Taiwan, Ho Chi Minh City, Jakarta, Kuala Lumper, Cairo, and Tehran, have evolved into automobile-saturated cities (Newman and Kenworthy 2015:95). Unfortunately, around the world, the hegemony of automobility has penetrated the middle classes as a market of social status and modernity.

In developed societies roads and parking lots have displaced millions of hectares of farmland, including 16 million in the United States alone (Brown 2001:190–200). The competition between automobiles and croplands is one between rich and poor people around the world and governments which subsidize motor vehicle infrastructure are penalizing the poor to subsidize the automobility of the rich.

Conversely, many cities in developing societies are coming to resemble most North American cities characterized by a spatial separation between residences and work sites. Furthermore, as Bongardt et al. (2013:41) observe, "[s]hopping and leisure activities are undertaken in out-of-town areas. High fuel prices limit social exchange and work opportunities; many suburbs are run-down and develop into places of uncoordinated social unrest." Shiva (2008:51) reports: "The car has seriously divided India. People can no longer walk on the streets. Neighbors have turned into enemies over car parking. It has cut up rural India through land grabs for factories and highways."

For better or worse, people in many cities in the developing world travel on foot or bicycle as is indicated in table 3.4 below.

Benxi resembles the mode share of travel patterns of various African cities more than it does the major Chinese urban centers of Beijing and Shanghai whereas Mumbai as a booming Indian metropolis appears to approach the travel mode of Western cities. While a substantial number of people rely on cycling as their modal form of travel in Dar es Salaam, this is not the case in

Table 3.4. Urban Travel Mode Shares in Selected African and Asian Cities*

City	Country	Modal Form of Travel–% of Daily (Total Person Trips)			
		Walking	Bicycle	Public transportation	Private motor vehicles
Nairobi	Kenya	47	1	43	7
Addis Ababa	Ethiopia	70	1	24	2
Mumbai	India	13	10	23	60
Delhi	India	31	7	41	21
Beijing	China	34	38	16	12
Benxi	China	60	7	19	14

*Adapted from Pendakur (2011:213).

Nairobi and Addis Ababa. Indeed, according to Godard (2011:254), "cycling is not very common in African cities, with some exceptions," apparently because it is viewed as a mark of poverty. While railways had a tremendous impact on opening up remote parts of Africa under colonialism, beginning in the 1940s railways became less important as a mode of transport and have been "almost totally superseded by other means of transport," such as buses, trucks, automobiles, and motorcycles (Gewald, Luning, and van Walraven 2009:2). Thus, motorized vehicles often constitute the only public transport in many African cities.

In many cities in the developing world, pedestrians and cyclists often find themselves competing with motor vehicles for road space.

> In metropolitan cities such as Bangalore and Mumbai, the pedestrians are literally forced to walk on the roadway, thus risking their lives, simply because there are no pavements, because they are inadequate for the pedestrian volumes they serve and/or because they are discontinuous. In addition: bicycle lanes have too often given way to motor vehicles (as in the case of Beijing and Bangalore); pedestrian spaces like footpaths and foot bridges are occupied by hawkers and vendors (as in the case of Mumbai and Dar es Salaam); and motor vehicles are parked on footpaths and in bicycle lanes (a frequent occurrence in Beijing, Shanghai, and Mumbai). (Pendakur 2011:216)

In various large cities, elites have turned to helicopters to circumvent traffic congestion. Sao Paulo is the most extreme example of this trend. It is not only the largest city in South America but has "what is probably the world's heaviest urban helicopter traffic" (Cwerner 2006: 192). While helicopters for the time being may be a convenient form of mobility for avoiding congestion, they are expensive, and energy-intensive, thus contributing to air pollution and greenhouse gas emissions, very noisy, and extremely dangerous. With over 250 heli-pads, Sao Paulo is the "city with the largest number of elevated helipads anywhere in the world (with probably ten times as many elevated helipads as any other major city" (Cwerner 2006:206).

CONCLUSION

This chapter discuss the initial role of public transportation in shaping suburbanization and land transformation, continuing with the impact of the rise of the motor vehicle on these processes. It also examines the rise of congestion resulting from large-scale motorization in both developed and developing societies. Motor vehicles have become seemingly omnipresent in cities, towns, and even villages around the world, whether they are moving or parked.

While certainly trains and trams require a large amount of land, motor vehicles do so much more. Particularly the megacities of the world are literally being choked by the presence of motor vehicles of various sorts. In various cities in developing countries, the presence of motorbikes exceeds that of automobiles. As Hansen (2017:641) observes, the motorbike constitutes a "crucial part of everyday mobility for a large portion of the world's inhabitants." Driven by the needs of capital and by population growth, often a by-product of poverty, cities around the world are growing, generally accompanied by more motor vehicles, roads, and parking spaces. So long as the culture of automobility remains hegemonic, these trends will continue. In many cities around the world, private motor vehicles have displaced public transit in the form of trains, trams, and buses. Some developing countries have limited access to train lines, particularly in South America and Africa, and rely upon automobiles, buses, motorcycles, mopeds, in urban areas and for connecting different towns and cities.

Down under, politicians affiliated with the two major parties, the Coalition and the Australian Labor Party (ALP), repeatedly speak of plans to increase the population of the metropolitan area from 4.6 million people to eight million by 2050, boasting that in the foreseeable future that Melbourne's population will exceed that of Sydney, which presently stands at about five million people. Occasionally there is discussion of diverting Australia's population growth, largely driven by immigration, to regional centers with accompanying infrastructure and educational, health, recreational, and transportation amenities, but generally such discussions are sidelined. However, as we will see in subsequent chapters, there is now increasing discussion on how to make cities less automobile dependent and more environmentally and socially liveable, particularly in developed countries but to some extent in developing countries. However, in terms of cities around the world, Melbourne and Sydney, with a population of some five million people each, remain much smaller than the growing number of megacities with populations over 10 million people and in some instances over 20 million people. While the topic lies beyond the purview of this book, we might pose the question, what would a liveable and sustainable city look like in terms of population, density, and transportation? In terms of transportation, how might alternative forms of mobility such as walking, cycling, trains of various types, foster the creation of more compact and liveable cities that avoid the sprawl that the automobile has fostered and continues to foster around the world? Yet, other challenges in terms of settlement patterns are how should the space between cities and even smaller communities, namely small towns and villages, be organized and where do spaces for agriculture, forestry, extensive parklands and wilderness areas, fit into the equation?

Chapter Four

The Impact of Motor Vehicles on the Natural Environment

In addition to motor vehicles having a profound impact on settlement patterns or the social environment so to speak, they also have a profound impact upon the natural environment, particularly the atmosphere which provides us with oxygen essential to breathing and plays an essential role in maintaining a stable eco-system. Unfortunately, motor vehicles emissions emit sulphur oxides, nitrogen oxides, carbon monoxide, hydrocarbon, volatile organic compounds, toxic metals, lead particles and particulate matter, and carbon dioxide, which contribute to acid rain, climate change, and cardiovascular, pulmonary, and respiratory ailments. Anthropologist Lenora Bohren (2009:370) posits the car as a form of cultural adaptation that became highly maladaptive in that it "has caused many unintended consequences to the environment." She notes that the "car also brought on problems of intense resource and energy use and environmental degradation" (Bohren 2009:372).

THE CONTRIBUTION OF MOTOR VEHICLE PRODUCTION TO THE DEPLETION OF NATURAL RESOURCES AND ENVIRONMENTAL DEGRADATION

Motor vehicles are part and parcel of the metabolic rift that capitalism and post-revolutionary societies create with nature. Like other consumer goods, private motor vehicles touch the natural world at four points: (1) extraction of raw materials for their production; (2) their manufacture which contributes to environmental pollution and greenhouse gas emissions; (3) consumer use which may further contribute to more environmental pollution; and (4) their disposal once they have outlived their usefulness (McCarthy 2007: xiii).

Motor vehicle production entails energy expenditure, scientific knowledge, continual technological innovation, electronics, computer programming, sophisticated manufacturing procedures, and a considerable amount of shipping back and forth across the globe of component parts. Sometime ago, Peter Freund and George Martin (1993:15) asserted that the automobile "contributes to worldwide resource depletion." Auto production requires tremendous quantities of copper, steel, and light metals, particularly aluminium. Foster observes:

> [F]ully, one-third of the total environmental damage caused by automobiles occurred before they were sold and driven. One study estimated that manufacturing the average car produced 29 tons of waste and 1,207 million cubic yards of poisoned air. This information is significant because it calls into question the notion that keeping exhaust-spewing old vehicles, or "beaters," running is worse for the environment than trading them in for new automobiles every three years or so. (Foster 2003:132–133)

Over 50 percent of the world's rubber plantations provided the raw material for original and replacement tires for automobiles and trucks until World War II when synthetic rubber derived from petroleum began to serve this purpose (Penna 2015:229). Catalytic converters, and fuel cells use platinum and electric vehicle batteries require lithium, most of which comes from Bolivia.

Motor vehicle production at the Ford Rouge plant, particularly during the 1950s and 1960s, resulted in high levels of air and water pollution, which spurred residents in the South End of Dearborn to protest the former, and sportsmen to protest against the latter (McCarthy 2007: xviii). The production of automobiles at the Rouge plant required massive amounts of pig iron from processed from ore, limestone, and coke, which left blast furnace slag as debris, prompting Ford to construct a cement plant on-site to eliminate the expense of disposing some 125 tons of slag each day (McCarthy 2007:69).

In the mid-1990s, the Environment and Forecasting Institute in Heidelberg calculated the resources required in the manufacture, operation, and disposal of a prototypical German middle-class automobile over a ten-year period for which they found the following sobering facts:

- Its manufacture produced 29 tons of waste and 1,207 million cubic yards of air pollution
- Its operation resulted in 1,330 million cubic yards of additional air pollution
- Its disposal resulted in 66 tons of CO_2 and 2.7 billion cubic tons yards of polluted air (Kay 1997:92–93)

In the 1990s, most of the waste associated with automobiles came from the manufacturing process, with some 25 tons of waste coming from the production of each car (Graves-Brown 1997:65).

In terms of locomotion, motor vehicles generally rely heavily upon petroleum or oil, which when burned, emits carbon dioxide, an issue that is given greater focus in the next chapter. While oil initially was used for lamp kerosene, which quickly was displaced by electricity for illumination, by the late nineteenth century oil had evolved into the primary fuel of the internal combustion engine so crucial to the evolution of automobility. Oil, as John Urry (2016: 111) observes, "makes fast movement possible," not only in the case of motor vehicles but also diesel-locomotive-driven trains and ships, the latter of which ironically transport oil around the world.

In the United States, Texas, Oklahoma, and California particularly became major oil production sites during the early twentieth century. Oil reserves were shortly thereafter discovered in Russia, Eastern Europe, and the Middle East and eventually in many other locations around the world. In addition to automobiles, oil came to fuel trucks, train locomotives, ships, and airplanes. Oil replaced coal as the primary fossil fuel in the 1950s. Oil also became an essential ingredient in the production of plastics, which in large part replaced packing materials such as paper, wood, glass, and metal. Oil in many instances replaced coal in the heating of dwelling units, offices, and factories, but natural gas has also become a fuel used for this purpose.

Oil presently supplies about 40 percent of the world's energy and 96 percent of its transportation energy (Forest and Sousa 2006:1). Urry maintains:

> Oil makes fast movement possible. It provides almost all transport in the modern world (at least 95%), powering cars, trucks, planes, ships, and some trains. It thus makes possible the mobile lives of friendship, business life, professions, and families. Oil also moves components, commodities, and food around the world in trucks, planes, and vast container ships, as well as oil itself in large tankers. (Urry 2016:111)

Despite much discussion of the world purportedly approaching "peak oil," a hotly debated topic, the oil industry continues to seek new sources of oil, regardless of the environmental and climatic consequences. In the short run, it is now clear that the peak oil theorists did not anticipate rapid developments in technological innovations that are now driving production increases in new or previously abandoned locations. New sources of oil and fuel for motor vehicles, as well as trains and airplanes, increasingly come from offshore oil, tar sands oil, oil shale, liquefied coal, and biofuels such as ethanol and biodiesel.

Unfortunately, oil availability, even at rising prices, is likely to be an ongoing issue, forcing contributing to climate change for the foreseeable future. Conversely, while the price of oil increased dramatically between 2010 to mid–2010 to $115 per barrel,

> [I]international benchmark oil prices fell by well over 50% in 2015 and have remained in the $40–60/barrel range for much of 2015. The collapse in prices was driven by a marked slowdown in demand growth and record increases in supply, particularly tight oil from North America, as well as a decision by the Organization of Petroleum Exporting Countries (OPEC) countries not to try to rebalance the market through cuts in output. (International Energy Agency 2015:46)

Particularly due to the oil crises of the 1970s, some Americans opted for smaller cars, such as the Volkswagen (which had already become popular) and Japanese cars, which initially became popular on the West Coast and spread to other regions of the country. Obviously smaller automobiles were more environmentally friendly than larger ones in that they "burned less gasoline, which meant less smog-causing emissions" and "required smaller quantities of raw materials, entailed less manufacturing pollution, and left less material to dispose at the end of their lives" (McCarthy 2007:214). Conversely, they generally had to be imported from far-off places, such as Europe and East Asia, which has contributed to undetermined fuel expenditure, air and water pollution, and greenhouse gas emissions due to shipping. In the 1980s, during the Reagan era, a large number of Americans however turned to SUVs, which are more environmentally destructive and require more resources to manufacture than small automobiles. Reportedly SUV sales increased from 132,000 per annum to 800,000 per annum between 1982 and 1985 alone and high sales of SUVs continued into the twenty-first century, not only in North America but other countries around the world (Penna 2015: 229). Road construction, which is an integral part of motor vehicle transportation, relies upon the extraction of limestone, sand, crushed stone, coal tar, and oil to produce concrete and asphalt (Penna 2015:229). Motor vehicles also require inordinate amounts of rubber. China's motor vehicle fleet began to dramatically increase after 2000 and extracted much of its rubber from the Mekong River region in its southwest (McNeill and Engelke 2014:159).

Historically automobile companies have pressured the state against environmental regulation. They have supported national lobby groups such as Coalition for Vehicle Choice in the United States and international lobby groups such as Global Climate Coalition and Climate Council (Mikler 2009:6).

Given over the past two decades or two recognition of the ecological and climate crises has grown in the public consciousness, automobile companies have turned to a form of "corporate environmentalism" and support and belong to industry groups such as the World Business Council for Sustainable Development. They have even formed linkages with some environmental NGOs.

AIR POLLUTION

Along with industrial pollution, motor vehicles have contributed to air pollution in cities around the world, particularly ones in developing countries. CBDs and inner-city areas tend to be the most adversely impacted by motor vehicle pollution due to the concentration of motor vehicles in them, often slowing inching along in congestion, and spewing out exhaust fumes. Ironically, Keucheyan (2016:29) observes, "[T]hese are areas where a significant portion of the bourgeoisies of Europe live [as well as other regions, both in the developed and developing countries]. City centres also tend to be less well supplied with green spaces than certain suburbs are."

In terms of developed countries, the Los Angeles metropolitan area has been long recognized for a high level of motor vehicle-induced air pollution. However, it took a while for experts, politicians, and the public to come to terms with the role of motor vehicles in contributing to air pollution and more specifically smog. Perhaps the triggering incident in bringing about this awareness occurred in the last ten days of July 1943 when Los Angeles experienced its first major smog episode in which a "brownish blue haze that greatly reduced visibility and a stinging sensation in the eyes that caused tearing" (McCarthy 2007:116). Although industrial activities in the area contributed to air pollution, motor vehicles constituted the primary cause, a stance that Arie J. Haagen-Smith, a plant biochemist at the California Institute of Technology adopted in late 1947 in his role as a consultant to the Los Angeles Air Pollution Control District (APCD). In 1950, Haagen-Smit in a lecture "announced that Los Angeles smog was caused by a chemical reaction involving unburned hydrocarbons and nitrogen oxides in the atmosphere that produced ozone" and that oil refineries and automobiles were the primary sources of these pollutants (McCarthy 2007:118). Around the globe, leaded and unleaded motor vehicle emissions constitute the greatest contributor to ground-level ozone, which contributes to smog, haze, and various respiratory complications, including coughing, sneezing, shortness of breath, and asthma (Penna 2015:297).

In the United States, motor vehicle-induced air pollution worsened during the 1960s. Doyle reports:

> In 1964, automobiles accounted for an estimated 430 tons per day of NOx; by 1968 it was up to 645 tons per day. By 1968, Los Angeles County was recording a third to a half of its calendar year had pollutants at the "adverse level"—132 days in excess of the California nitrogen dioxide standard, then set at 0.25 PPM for one hour, 188 days in excess of the state's 0.15 PPM oxidant (ozone) standard. (Doyle 2000:25)

Despite improvements in US automobile fuel economy and emissions controls standards, a doubling of miles driven during the 1980s and 1990s negated the impact of these innovations. Furthermore, while the catalytic converter "effectively breaks down the various nitrous oxides [NO_x] that contribute to smog and local air pollution . . . it creates nitrous oxide [N_2O], benign smog creation but 300 times more potent than carbon dioxide as a greenhouse gas" (Porter 1999:81). The US Bureau of Transport Statistics (2009) reports that as of 2007, mobile forms of transportation accounted for 3.7 percent of sulphur dioxide emissions, 57 percent of nitrous oxide emissions, 68.4 percent of carbon monoxide emissions, 2.9 percent of PM10 particulates, 18.8 percent of PM2 particulate emissions, and 33.9 percent of volatile compound emissions as well as a substantial portion of ozone emissions in urban areas (Black 2010:7). Particularly problematic are idling, all too often in congestion, with stop-and-go traffic, and decelerating. The former "emits significant amounts of pollutants in the form of emissions such as carbon monoxide and hydrocarbons" and the latter means that "fuels in the process of being combusted are suddenly expended without creating the energy that they would have," thus producing large amounts of carbon monoxide and hyrdocarbons" (Black 2010:162). Despite the implementation of motor vehicle emissions standards in the United States, over half of Americans reside in "cities that do not consistently meet the federal standard and more than 80 million live in cities that consistently fail the standard" (Penna 2015:297).

Mage and Zali (1992: viii) report: "Exposure to high levels of motor vehicle pollutants occurs essentially in three situations: a) while inside vehicles (from immediately surrounding traffic); b) while working in, or walking alongside congested areas, and c) through residence in urban neighborhoods with high motor vehicle traffic pollution (and or ozone, in urban areas downwind from the city center." Of the estimated 4.4 million tons of human-generated pollutants emitted into the air of Mexico City in 1989, 76 percent were produced by motor vehicles (Freund and Martin 1993:67). In contrast, of the 3.5 million tons of human-generated pollutants emitted into the air of Los Angeles—perhaps America's most polluted city—in 1985, 63 percent were created by motor vehicles. Sometime ago, Sweezy (1973:4) compared auto-generated pollution and traffic congestion to the "outward symptoms of a disease with deep roots in the organs of the body." In essence the automobile has become a major form of assault on the social and ecological body. Continuing degradation of air quality prompted the Union of Concerned Scientists in 1991 to shift its primary concern from nuclear energy to the internal combustion engine and the National Resources Defense Council to declare the following year that the automobile was "the worst environmental threat in many U.S. cities" (Kay 1997:80).

Cities in developing societies particularly are adversely subjected by air pollution, with 65–70 percent of it emanating from motor vehicles (Halder 2006:28). Air quality standards in most developing countries constitute a lower priority than provision of clean drinking water, sanitation, housing, healthcare, personal safety, and abject poverty. Bohren (2009:374), views the growth of car culture as an important force in making Mexico City the "most populated and polluted city in the world." The level of air pollution or suspended particulate matter (SPM with the figure expressed in micro gram/cubic meter) in the five most polluted cities in India in 1997 was as follows, with most of it coming from motor vehicles: 460 in Delhi, 460 in Calcutta, 350 in Kanpur, 230 in Nagpur, and 230 in Jaipur (Halder 2006:29). The permissible limit specified by the World Health Organization is 200 SPM.

As table 4.1 below indicates, environmental regulations in the United States resulted in significant reductions overall in emissions from motor vehicles.

Table 4.1. Emissions from Motor Vehicles in the United States, 1970 and 2002

Type of vehicle and pollutant	Emissions in grams/kilometer from all vehicles on the road			
	1970	2002	Change (%)	Change in total emissions (%)
Light-weight vehicles				
Carbon monoxide	76.4	12.5	−84	−59
Nitrous oxides		5.45	0.76	−86
Volatile organic compounds		7.98	0.88	−89
Particulate matter		0.17	0.02	−90
Sulfur dioxide		0.08	0.03	−59
Heavy-duty vehicles				
Carbon monoxide		197.8	9.4	−95
Nitrogen oxides		23.16	9.98	−57
Volatile organic compounds		19.38	1.07	−94
Particulate matter		1.43	0.32	−78
Sulfur dioxide		1.09	0.31	−72

Source: Adapted from Gilbert and Perl (2010:190).

Although the emissions per kilometer from light-duty vehicles fell substantially between 1970 and 2002, the total distance traveled by them increased from 1.68 to 4.30 trillion kilometers (Gilbert and Perl 2010:190). There was an increase in distances traveled for heavy-duty vehicles from 100 to 345 billion kilometers between 1970 and 2002, thus more than offsetting the reduction in NOx emitted per kilometer and thus resulting in an increase in NOx emissions from heavy-duty vehicles (Gilbert and Perl 2010:191).

Table 4.2 below indicates motor vehicle emissions per capita in the 25 EU countries were appreciably lower than in the United States in 2002, largely because "Europeans use only about a third as much transport fuel per capita" (Gilbert and Perl 2010:197).

Bongardt et al. (2013:38) report that "[e]xposure to such emissions can lead to increased blood pressure, liver/kidney damage, impairment of fertility, comas, convulsions, and, in the worse cases, death." One of the major by-products of gasoline exhaust is benzopyrene, a carcinogenic chemical that is suspended in urban air. Bowdon (2004:16) reports: "When released into the atmosphere, even in relatively small quantities, it can cause lead poisoning in humans. The effects of lead poisoning can include headaches, stomach pains, tremors and in severe cases even death. Lead is especially harmful to young children as it affects the development of the brain."

Although lead has been gradually removed from motor vehicle fuel in developed countries since the 1970s, it is still present in fuel in many developing countries. The high levels of carbon monoxide coming primarily from motor-vehicle emissions in congested areas "can lead to levels of 3% carboxhhemoglobin (COHB) which produce adverse cardiovascular and neurobehavioral effects and seriously aggravate the condition of individuals with ischemic heart disease" (Mage and Zali 1992:vii). Another motor vehicle

Table 4.2. Road Transport Emissions Per Capita and Energy Use Per Capita, US and EU25, 2002

	Kilograms per capita		
	US	EU25	EU25/US (%)
Carbon monoxide	128.9	42.4	33
Nitrogen oxides	24.7	11.7	48
Volatile organic compound	24.3	8.1	33
Particulate matter	1.3	0.9	68
Sulfur oxides	2.6	0.3	12
Road transport energy use (Giga Joule/person)	79.2	26.0	33

Source: Adapted from Gilbert and Perl (2010:198).

emission gas, nitrogen dioxide, produces a brownish-red haze and can contribute to respiratory problems, particularly in asthmatics and young children.

The American Lung Association estimated that in 1985 motor vehicle pollution contributed to some 120,000 deaths in the United States (Freund and Martin 1993:29). Some brake linings contain asbestos, a well-known carcinogen. Motor vehicle emissions contribute to elevation of ozone levels, of which Greater Atlanta is a prime example. The city has evolved into a leading US metropolitan area in vehicle miles traveled per person per day at 34.1, as opposed to other leaders, such as Dallas (30.1); Washington, DC (22.6); and Los Angeles (21.5). The result is massive air pollution, including ozone, which contributes to various respiratory complications, including asthma. Research Atlanta, an institute affiliated with Georgia State University, has reported that asthma-related visits to the pediatric emergency clinic at Grady Memorial Hospital increased by a third during high ozone days (Doyle 2000:3). The Centers for Disease Control and Prevention estimates that the number of asthmatics in the United States increased from 6.8 million in 1980 to 17.3 million in 1998, in part due to ozone pollution (Doyle 2000:4). An estimated 100 million Americans live in places that do not meet the EPA's eight-hour-air-quality standard for ozone (Doyle 2000:5).

The Clinton administration's creation of a Partnership for a New Generation of Vehicles in the 1990s provided funding to national laboratories and auto-parts companies with the objective of creating a mid-sized automobile that would get 80 miles to the gallon, but the growing popularity of gas-guzzling, expensive sports utility vehicles (SUVs)—which the US government classifies as "light trucks," and so are subject to less stringent emissions standards—significantly added to air pollution during this period (Bradsher 2002:70). Ironically, SUVs particularly appeal to baby boomers, many of whom view themselves as sensitive to environmental issues, as well as movie stars and directors, singers, and other entertainment idols. At the other end of social ladder, people of low socioeconomic status who do not even own cars are those at greatest risk for traffic-related air pollution, again because they tend to live relatively close to major thoroughfares. While as Black (2006:262) observes, it is difficult to precisely determine how much of air pollution is attributable to motor vehicles, he reports based on research in the Los Angeles area: "Transportation sources account for some 60 to 90 percent of all air pollution in Los Angeles County and three adjacent counties" and that in California, "all forms of air pollution cause some nine thousand premature deaths, nine thousand hospitalizations, 1.7 million cases of respiratory illness, 1.3 million school absences, and 2.8 million lost days, according to the state's Air Resources Board."

While developing countries generally have relatively few cars per 1,000 population, "each vehicle is likely to emit more air pollutants per mile than the vehicles in developed nations due to a lack of or less stringent emission controls, and/or poor quality," particularly adversely impacting upon informal workers, such as street vendors and hawkers, who are more prevalent in developing countries (Flachsbart 1992:106). Ray and Lahiri identify three significant sources of air pollution in Indian cities:

(i) emissions from motor vehicles due to combustion and evaporation of automotive fuels,
(ii) emissions from industrial units and construction of buildings, infrastructure in and around cities, and
(iii) emissions from domestic sources (Ray and Lahiri 2010:167)

While it is difficult to determine precisely the health consequences of each of these sources of air pollution, they note that of these emissions, the contribution of vehicular emissions has been increasing over the years whereas "emissions from industrial and domestic sources are declining because coal-based industries are increasingly using cleaner fuel, highly polluting industrial units are being shifted outside the city areas, and liquefied petroleum gas (LPG), a relatively cleaner fuel, is replacing coal and biomass as the cooking fuel in urban households" (Ray and Lahiri 2010:167–168). Ray and Lahiri (2010:169) report that the number of motor vehicles in India increased by a factor of 29, from 1.9 million in 1971 to 55.0 million in 2001, over the course of three decades. Sixty percent of the residents of Calcutta, India, were found to have pollution-related respiratory problems (Freund and Martin 1993:67). An increase in the number of registered motor vehicles from 281,687 in 1980 to 785,352 in 1998 contributed to an increase in air pollution in Mumbai and the "emission of various pollutants from urban transportation is expected to grow 3–5 times in the next 20 years" (Yedla, Shrestha, and Anandarajah 2005: 246).

In a similar vein, the number of motor vehicles in Brazil doubled during the period of 1999–2009, reaching some 59 million vehicles in 2009, resulting in increased levels of atmospheric air pollution through the release of carbon monoxide and other harmful particulates resulting in various respiratory problems and allergies (Assis and Silva 2012: 2169). The number of automobiles in Mexico City skyrocketed from some 100,000 in 1950 to four million by 1990, resulting in a nearly constant haze. McNeill and Engelke report:

> Trucks, buses, and cars accounted for 85 percent of Mexico's air pollution, which by 1985 was occasionally so acute the birds fell from the sky in mid-

flight over the central square (the Zocalo). After careful monitoring began in 1986, it emerged that Mexico City exceeded legal limits for one or more major pollutants more than 90 percent of the time. In the 1990s, estimates suggested some six thousand to twelve thousand annual deaths were attributable to air pollution in the city, four to eight times the annual number of murders. Various efforts to curb air pollution since the 1980s have produced mixed results, but the death rate seems to have declined slightly since the early 1990s. (McNeill and Engelke 2014:24–25)

In the period 2000–2006 emissions were reduced about ten percent due to various measures, such as replacement of old catalytic converters in automobiles, reduced traffic, improvement of automobile pollution inspections, modernization of public transportation, adoption of rapid bus transportation, along with improved industrial pollution control (Scheinbaum 2009:2009). Nevertheless, heavy reliance on a massive taxi fleet and "ad hoc *collective* buses and *combi* minivans, rather than on public bus routes or high-quality express transit systems" continues to contribute to a serious air pollution problem in Mexico City (Calthorpe 2016:103).

Air pollution from motor vehicles has also become a serious problem in cities in developing countries, such as Jakarta, Beijing, Sao Paulo, and Addis Ababa, during the last decades of the twentieth-century (Etyemezian et al. 2005). Motor vehicles have been a major contributory factor to air pollution in Beijing, along with manufacturing, making it one of the most polluted cities in the world. Calthorpe (2016:95) reports: "Merely to ensure blue skies during the 2008 Olympics, the city spent some $17 billion restricting traffic and shutting down factories. It even employed 50,000 people to fire silver iodide at clouds to release rain."

Traffic congestion has become an increasing problem around the world, particularly in developing countries. In addition to increasing commuting time and fuel consumption, frequent stop-and-go driving results in more vehicular pollutants, including carbon monoxide, hydrocarbon (HC), oxides of nitrogen (NO_x), and fine particular matter (PM) and carbon dioxide, a greenhouse gas. One study found that in Indian cities automobiles consume 10 percent more fuel and emit 20 percent more carbon monoxide, HC, and NO_x during rush traffic than non-rush hour traffic (De Vlieger, De Keukeleere, and Kretzschmar 2000). Auto-rickshaws (a popular three-wheeler hire-transportation modality) in most Asian cities with the two-stroke engines consuming about 20 percent more fuel and emitting more CO_2 than the more expensive four-stroke engines (Choudary and Gokhale 2016:59). India reportedly has about three-quarters of the auto rickshaws in the world (Harding et al. 2016: 143). The Environment Pollution (Prevention & Control) authority for the National Capital Region of India, namely Delhi, maintained that the some 80,000 auto-rickshaws existing in the city in 1999 significantly contributed to urban air pollution due to their two-stroke engines and intensive use and ordered their drivers, along with taxi cab drivers, to replace the older vehicles with new vehicles that conformed to emissions standards, but the extent to which this mandate was carried out is unclear (Harding et al. 2016: 145–146).

GREENHOUSE GAS EMISSIONS

The 2014 Intergovernmental Panel on Climate Change states:

> The transport sector produced 7.0 $GtCO_2$ eq of direct GHG emissions (including non-CO_2 gases) in 2010 and hence was responsible for approximately 23% of total energy-related CO_2 emissions (6.7 $GtCO_2$). Growth in GHG emissions has continued since the Fourth Assessment Report (AR4) in spite of more efficient vehicles (road, rail, water craft, and aircraft) and policies being adopted. (Sims 2014:603)

The IEA/OECD (2012) reports that transportation accounts for 22 percent of CO_2 emissions, which it projects will increase by 35 percent by 2035. The European Conference of Ministers of Transport (2007:5) reports that "[t]ransport sector emissions grew 1,412 million tonnes (31%) worldwide between 1990 and 2003, and increased 820 million tonnes (26%) in OECD countries" and that the "OECD-ECMT region accounts for 71% of worldwide CO_2 emissions from transport." The ECMT countries are Albania, Bulgaria, Malta, Romania, Armenia, Azerbaijan, Belarus, Estonia, Georgia, Latvia, Lithuania, Moldova, Russia, Ukraine, Bosnia Herzegovina, Croatia, FYR Macedonia, Serbia and Montenegro, and Slovenia. Reportedly,

> The USA, Japan and China regulate passenger car fuel efficiency, and Japan also regulates heavy duty vehicle fuel economy. The EU and its Members States together with Switzerland, Australia and Canada all employ voluntary targets for car manufacturers and importers. Japan has by far the most ambitious regulatory standards, but the EU voluntary targets are of a similar order. US standards are far less ambitious, with the exception of the new standards adopted by California in 2006. (European Conference of Ministers of Transport 2007: 15)

A more recent OECD document (International Transport Forum 2010) reports the following statistics:

- "Transport-sector CO_2 emissions represent 23% (globally) and 30% (OECD) of overall CO_2 emissions from fossil fuel combustion"
- "Global CO_2 emissions from transport have grown by 45% from 1990 to 2007"
- Road freight accounts up to 30–40 percent of road sector CO_2 emissions
- Emissions from global aviation and international shipping accounted for 2.5 percent and 3 percent, respectively, of total CO_2 emissions in 2007
- Selected countries, particularly France, Germany, and Japan, experienced a stabilization or decrease in their road CO_2 emissions prior to the economic crisis of 2008–2009

In that they emit greenhouse gases, particularly carbon dioxide and nitrous oxide, as well a black carbon, motor vehicles are a major contributor to climate change and its various health impacts. Given that oil is a fossil fuel and thereby a contributor to greenhouse gas emissions, we must bear in mind that the products of the global automobile industry account for nearly half global oil consumption (Dauvergne 2008:43). The car consumes up to 63 percent of the oil used in the United States and about 35 percent of the oil consumed in Japan, a country with a vastly superior public transport system. Oil is also a major resource utilized in road construction (Paterson 2007:38). Automobiles require an inordinate amount of fuel, a demand that has been spurred on by the growing demand in developed societies for bigger and bigger vehicles. For example, the sports utility vehicle (SUV) or four-wheel drive market share in the United States increased from 2 percent in 1975 to 24 percent in 2003 (Leggett 2005:22). While it is often recognized that transportation, including motor vehicles, is a major contributor to greenhouse gas emissions, figures on the sector's contribution "typically do not include wider environmental costs caused by the exploration, extraction, conversion and transport of primary energy carriers required to provide the final energy for the vehicles" (Reichert, Holz-Rau, and Scheiner 2016:25).

The United States Environmental Protection Agency (2015b:1) reported that transportation accounted for 28 percent of US greenhouse gas emissions in 2012. Light duty vehicles contributed 62 percent of these emissions, medium-and heavy-duty trucks 22 percent, aircraft eight percent, rail two percent, ships and boats two percent, and other modes of transport four percent. Table 4.3 depicts the increase of greenhouse gas emissions from on-road vehicles in 1990 and 2012, respectively.

Passenger cars are defined as "automobiles used primarily to transport 12 people or less" and light-duty trucks are defined as "vehicles used primarily

Table 4.3. Increase of Greenhouse Emissions from On-Road Vehicles, 1990–2012 US On-Vehicle Transportation Emissions (Tg–CO_2 equivalent)

	1990	2012	% Change 1990–2012
On-road vehicles	1,236.2	1,558.4	26.2
Passenger cars	657.4	793.8	20.8
Light-duty trucks	336.6	338.4	0.5
Motorcycles	1.8	4.3	141.1
Buses	8.4	18.6	122.2
Medium- and heavy-duty	231.1	403.4	74.5

Source: Adapted from United States Environmental Protection Agency (2015a:3).

for transporting light-weight cargo or which are equipped with special features such as four-wheel drive for off-road operation" (United States Environmental Protection Agency (2015a:4). This category includes SUVs and minivans.

The Environmental Defense Fund (2007) provides the following sobering statistics on motor vehicles and their contribution to greenhouse gas emissions in the United States alone:

- 30 percent of the world's automobiles are situated in the United States.
- There are 232 million registered vehicles.
- US cars and light trucks traveled 2.7 trillion miles in 2004.
- The average US car consumes 600 gallons of gasoline per year.
- The average US car emits 12,000 pounds of CO_2 each year (for a total of 2,784 trillion pounds of CO_2/year).
- The United States accounts for 45 percent of the world's automotive CO_2 emissions.

Internal combustion engines reportedly waste more than two-thirds of the fuel that they burn and emit 20 pounds of CO_2 into the atmosphere for each gallon of gasoline burned (Sperling and Gordon 2010:18). Despite the fact that Eckersley (2004:279) dubbed Sweden as a working prototype of the "green state," it reportedly overall has the highest-pollution-emitting cars, particularly Volvos and Saabs, in Western Europe.

Aside from the amount of CO_2 that cars emit, CO_2 emissions "required to produce the energy to make the car in the first place as well as the energy to make and maintain the roads and to extract the fuel and raw materials that go into the car's tank, tires, and body" also contribute to emissions (Robbins, Hintz, and Moore 2010:136). Cities vary greatly in terms of CO_2 emissions as well as other pollutants.

The US Environmental Protection Agency reports:

> The amount of CO_2 creating from burning one gallon of gasoline depends on the amount of carbon in the fuel. Typically, more than 99% of the carbon in a fuel is emitted as CO_2 when the fuel is burned. Very small amounts are emitted as hydrocarbons and Carbon monoxide, which are converted to CO_2 relatively quickly in the atmosphere. ... The EPA and other agencies use the following carbon content values to estimate CO_2 emissions. CO_2 emissions from a gallon of gasoline 8,887 grams CO_2/gallon CO_2 emissions from a gallon of diesel 10,180 grams CO_2/ gallon. Diesel creates about 15% more CO_2 per gallon. However, many vehicles that use diesel fuel achieve higher fuel economy than similar vehicles that use gasoline, which generally offsets the higher carbon content of the diesel fuel. (United States Environmental Protection Agency 2014:1)

The US EPA estimates that the typical passenger vehicle emits about 411 grams of CO_2 per mile. Gasoline-powered vehicles average about 21.6 miles per gallon and travel 11,400 miles per year and emit 4.7 metric tons of CO_2 in the process (EPA 2014:2). Furthermore, motor vehicles emit methane and nitrous oxide from the tailpipe and hydrocarbons (HFC) from leaking air conditioners, but unfortunately it is difficult to estimate their amounts (United States Environmental Protection Agency 2014:2–3).

The amount of greenhouse gases generated by energy sources that may fuel motor vehicles varies. The US EPA (2015:1) reports that 27 percent of US greenhouse gas emissions are generated by transportation, as opposed to 31 percent by electricity generation, 21 percent by industrial processes, 9 percent by agriculture activities, 6 percent by commercial activities, and 6 percent by residences. In the case of the US transportation emissions, 60 percent come from light duty vehicles (passenger cars, SUVs, and pick-up trucks), 23 percent from medium- and heavy-duty trucks, 8 percent from aircraft, 2 percent from ships and boats, 2 percent from rail, and 4 percent from other sources (US Environmental Protection Agency 2015:2). Kahn Ribiero et al. (2007) report that about one quarter of world energy-related greenhouse gas emissions can be attributed to transportation and nearly 85 percent of transportation-related greenhouse gas emissions are caused by land modes of transportation. NASA reports:

> Cars, buses, and trucks release pollutants and greenhouse gases that promote warming, while emitting few aerosols that counteract it. In contrast, the industrial and power sectors release many of the same gases—with a larger contribution to radiative forcing—but they also emit sulfates and other aerosols that cause cooling by reflecting light and altering clouds. (quoted in Shahan 2010:1)

While Seattle, Montreal, and Curitiba rely on public transportation more than cities in the US, Canada, and Brazil, respectively, road transportation (passenger and freight) accounted for 64 percent of greenhouse gas emissions in Seattle in 2012; 39 percent of greenhouse gas emissions in Montreal in 2009; and 61 percent of greenhouse emissions in Curitiba in 2016 (Mercier et al. 2016).

Trains, buses, and trams produce greenhouse gases as well as other pollutants, even if the power propelling them is obtained from coal-fired power plants (Black 2010:183). Nevertheless, overall cities in which public transportation coupled with walking and cycling result in considerably few greenhouse gas emissions per capita than those that are automobile dominant. Cities and countries vary greatly in terms of CO_2 emissions. Whereas the transport-related CO_2 in the New York metropolitan area totaled 3,378 kilograms per capita in 1990, it was 5,193 kilograms per capita in the Houston area (Newman and Kenworthy 1999:120). By contrast, Toronto has 46 percent less CO_2 per capita production than the average US city, largely due to an extensive public transportation system. While the catalytic converter "effectively breaks down the various nitrous oxides that contribute to smog and local air pollution . . . it creates nitrous oxide, benign in smog creation but 3000 times more potent than carbon dioxide as a greenhouse gas" (Porter 1999:81). According to Shiva (2008:55), "Although the Nano gets good gas mileage, it is a fossil-fuel driven car [which] at 1 million cars a year [will] contribute heavily to greenhouse gas emissions." Urban transportation accounts for about half of Bangkok's total energy use and CO_2 emissions (Schaefer 2011:113).

Extraction of motor vehicle fuel from non-conventional sources, such as the tar sand pits of Northern Alberta, oil shale in the Rocky Mountain states, and liquefied US coal promise to "hyper-carbonise the US transportation system," aside of rhetoric on the part of the Obama administration to decarbonize the US economy (Gordon and Burwell 2013:24). Furthermore, cement and concrete are important components of not only building materials but also road construction, parking lots, and even sidewalks, footpaths, cycle lanes, and urban plazas and other public places. Chris Goodall (2007:229) reports: "Nobody is quite sure, but most estimates suggest that the world cement industry is responsible for about 5 percent of global emissions. This surprisingly large figure arises because the world is now producing about 2.5 billion tonnes of cement a year, almost a third of a tonne per person on the planet." The greenhouse emissions resulting from materials used in the construction and maintenance of roads is generally not counted into the contribution motor vehicles make to emissions overall.

Under modernization, greenhouse gas emissions from transportation of many sorts have dramatically increased in China over the past three decades or so. In term of freight transportation (which includes railway, highway, waterway, aircraft, and oil pipeline), greenhouse gas emissions increased from 163 million tons of CO_2e in 2000 to 978 million tons of CO_2e in 2011 (Tian et al. 2014:47–48). For highway freight alone, greenhouse gas emissions increased from 54 million tons of CO_2e in 2000 to 113 million tons of CO_2e in 2011.

FOOD PRODUCTION VS. BIOFUELS

There are four principal types of biofuels: (1) wood products and crop residues that are directly burned; (2) ethanol produced from sugars, starches, or cellulose; (3) biodiesel produced from oil crops or waste cooking oil; and (4) methane produced by natural gas, digestion of animal manures, or human sewage. The United States, Brazil, and Russia have developed large-scale biofuel or agro-fuel projects while China is ramping up in this area. In 2008, approximately 33.3 million hectares were devoted to the production of food-based biofuels and their coproducts, with the International Energy Agency projecting their production will rise 170 percent by 2020 (Fargione et al. 2010:351).

Although biofuels consumption is often touted as a climate change mitigation strategy, there is increasing evidence that for the most part they are not a panacea. Conventional or first-generation biofuels include ethanol, which is produced by fermenting sugar from corn (maize), sugarcane, wheat, sugar beet, or cassava; biodiesel produced from rapeseed, soybean, palm oil, or sunflower seed; and biogas produced from the anaerobic digestion of waste material, such as maize silage, organic wastes, and even manure. Advanced biofuels are still for the most part in the research and development stage and include cellulosic-ethanol consisting of various forms of lignocellulosic biomass; biomass-to-liquids diesel consisting again of various forms of lignocellulosic biomass; and algae-biofuels. In contrast to conventional biofuels, they can be produced on non-arable plots of land. Farigone et al. (2010:371) maintain: "Biofuels are the most land-intensive form of energy production. . . . The land requirements for biofuels have potential negative consequences for biodiversity and GHG emissions by causing, either directly or indirectly, the conversion of natural ecosystems to cropland."

While the "combustion of biofuels releases only the carbon taken up from the atmosphere by the plant matter during its growth," their production and distribution "can require substantial amounts of fossil fuel, and petroleum-derived fertilizer and pesticide, thereby releasing in net additions to atmospheric CO_2." (Gilbert and Perl 2010:137). Although conventional biofuels burn more cleanly than petroleum-based products, ethanol reportedly contributes to air pollution, including ground-level ozone, to the same extent as the latter. Furthermore, the production of palm oil to produce biodiesel largely for the European market has resulted in the cutting down and burning of forests in Indonesia and Malaysia, thus transferring sequestered CO_2 to the atmosphere.

Global biofuel production reportedly grew 22 percent between 2000 and 2010, but only 1.2 percent between 2010 and 2012 (IEA/OECD 2013). Biofuels account for about three percent of total road transportation fuel, but in Brazil they constitute 21 percent of the total road transportation consumption (Bongardt et al. 2013:74). Biofuel crop production requires a tremendous amount of water in terms of irrigation and processing, generally primarily the former. The production of biofuels from corn, sugarcane, soybean, palm fresh fruit bunches, rapeseed, and switchgrass also requires large amounts of nitrogen and phosphorus fertilizers (Fargione et al. 2010:368). Even the production of biofuels from residue matter, such as jatropha, a plant grown on marginal land in India and Africa is used in biodiesel production, but these areas frequently feed livestock. While corn ethanol has the most profound environmental impacts, not far behind are those of soybean biodiesel production. Both forms of biodiesel production result in significant soil erosion and use large amounts of herbicides.

Because conventional biofuel production requires large swaths of land, it has contributed to the increasing price of some crops, meaning that particularly poor people cannot afford them. Dennis and Urry (2009:68–69) delineate the major dilemmas associated with biofuel production:

- Increased shortage of agricultural crop land
- Increased deforestation, loss of biodiversity, and water pollution
- Increased food prices and food scarcity
- Domination of biofuels market by giant multi-national corporations, many of which already dominate agricultural production
- Some evidence indicates that more 1.29 gallons of oil are required to produce one gallon of ethanol

Magdoff states:

> The rising prices for all basic foods has, of course, hit the poorest countries the hardest—especially those that important significant quantities of their food—although the poor in every country in the world have been hurt. There has already

been food riots in many regions and concern has been expressed for the future stability of some thirty-three countries. (Magdoff 2008:42)

The United Nations reports that up to 60 million people may in the foreseeable future become agro-fuel refugees (Dennis and Urry 2009:68).

DISPOSAL OF MOTOR VEHICLES

The motor vehicle industry encourages people to continually purchase newer models and to dispose of their old ones, be it automobiles or pickup trucks. Nieuwenhuis states:

> The longer a product lasts, the less often it needs to be replaced and, therefore, the less often it needs to be produced, thus reducing overall production and resource use.... Although older cars pollute more than newer cars—thus generating an apparently greater environmental impact—this can largely be rectified by retrofitting of equipment. The mass car industry has long resisted the move towards more durable products with some exceptions. (Nieuwenhuis 2014:94)

The exceptions include luxury automobiles, such as Rolls Royces, Volvos, and Porsches, that are designed to last 20–30 years for their demanding customers. Antique or classic cars also demonstrate that cars may be renovated to last a long time, although again they tend to be a consumer item for generally affluent people. Members of the Federation of British Historic Vehicles owned 540,000 historic vehicles in 2004, "of which 406,000 were roadworthy and taxed" (Nieuwenhuis 2014:99). The European Union has some 1,950,000 historic vehicles. Poorer people in developed and developing countries often purchase second-hand automobiles.

Discarded motor vehicles are assembled in immense junkyards or graveyards around the planet. They have posed a monumental environmental and even health problem, such as breeding places for mosquitoes and rats, around the world. Automobile tires end up in huge dumps, some of which have caught fire. Many of them have ended up in junkyards where buyers may purchase needed parts for a reasonably low price. Conversely, in order to circumvent the expense of transporting used motor vehicles to junkyards, owners simply left them on their premises or even dumped them in a nearby empty lot.

The US steel industry during its zenith purchased auto scrap but stopped doing so, which "resulted in overflowing junkyards and the abandoned automobile problem of the 1960s (McCarthy 2007: xviii). While portions of discarded motor vehicles can be recycled, it constitutes a partial solution and

104   Chapter Four

may have served as a ruse for motor vehicle companies not to manufacture vehicles that last longer. While automobiles now have increased longevity, largely due to the use of plastics that do not rust, plastic parts "cannot be repaired or easily recycled" (Graves-Brown 1997:66). The increased replacement of mechanical components with electronic ones in automobiles have made it more expensive and difficult to repair them.

Tyler Durden (2014) has written a photographic essay about the stockpiling of unsold cars and maintains that "[t]here are literally thousands of these 'car parks' rammed full of unsold in practically every country on the planet." His aerial photographs depict thousands of unsold cars at Sheerness in the UK, many unsold cars parked at the Nissan Sunderland test track in the UK, 1,000s upon 1,000s of unsold cars on a runway near St. Petersburg, and various seaports with 1000s and 1000s of cars, many of which presumably will not be sold.

THE IMPACT OF OTHER FORMS OF TRANSPORT ON THE ENVIRONMENT: TRAINS, AIRPLANES, AND SHIPS

While motor vehicles obviously have negative impacts on the environment in terms of depletion of natural resources, air pollution, and greenhouse gases, so do other modes of transportation, including trains, airplanes, and ships of various sorts. For example, Vaze observes:

> [E]missions from a person driving alone in a car are three to five times worse than most public transport options. But with two people in the vehicle it is as good as the typical bus, with four people as good as a tube. Buses are significantly worse than the coach, tube and the train because of their low occupancy rates: the average bus carries 15 people despite having seating for 45 or more. The coach data from National Express [in the UK] rather cheekily assumes full occupancy—optimistic perhaps because most coaches do not have high occupancy. (Vaze 2009:131)

Aside of the energy consumed in the operating of these various modes of transportation, each of them requires a tremendous amount of energy consumption in terms of their manufacture and maintenance and the construction, maintenance, and operation of the infrastructure (train stations and airports) that support them.

Trains

Passenger and freight trains are generally depicted as less energy and greenhouse gas emissions-intensive than automobiles, trucks, and airplanes. However, during the era of steam locomotives this was not necessarily the case. According to Carpenter (1994:346), "[i]n the 1950s more pollutants were emitted by trains per passenger-km or per tonne-km of freight than petroleum-driven road vehicles." Steam locomotives still exist in parts of Eastern Europe, China, and other developing countries (Carpenter 1994:174). Even when steam locomotives are operating at full pressure, they are "only about 10 percent thermally efficient, compared with 20 per cent for petrol engines and 30 to 40 percent for diesel or electric traction" (Carpenter 1994:175). The Association of American Railroads, a body lobbying for greater reliance on rail freight over truck freight in the United States boldly asserts:

> According to a recent independent study for the Federal Railroad Administration, railroads on average are four times more fuel efficient than trucks. Greenhouse gas emissions are directly related to fuel consumption. That means that moving freight by rail instead of truck reduces greenhouse gas emissions by 75 percent. (Association of American Railroads 2008:1)

Passenger trains tend to be faster than freight trains. Emissions on trains increase when speed exceeds 200 km/hour (Bridger 2013:15). R. Kemp reports:

> Trains have always had a good reputation for energy consumption. A traditional 14 km h^{-1} diesel-hauled train of a locomotive and 8 coaches can carry 500 passengers and has fuel consumption of about 1 mile per gallon. The same passengers travelling 2 per car on a motorway, each car averaging 30 miles per gallon, would use 8 times more fuel. However, the relative benefit deteriorates as the train speed increases and the situation becomes more complicated to analyse if the train uses electricity generated by a mix of gas, coal and nuclear power. (Kemp 1995:77)

He maintains that rail transportation loses its energy consumption advantage over air transportation between 300 and 400 km/hour, but much of this is dependent on the aerodynamics and efficiency of specific trains or airplanes (Kemp 1995:79). Furthermore, in that sleeper cars generally carry fewer passengers than regular carriages, their emissions per passenger are higher than those who sit in the latter (Clark 2009:174).

Airplanes

Flying is reportedly the fastest-growing single source of greenhouse gas emissions and is expected to be so in the future (Bridger 2013:2). According to Eriksen (2016:59), "[g]lobal air traffic grew by 60 per cent between 2000 and 2012, in spite of temporary dips owing to the 9/11 attack, the SARS scare and the recession in 2007–9." After the German carrier Lufthansa pioneered civilian air travel in the 1920s, flying remained largely an elite mode of transportation until the 1950s. Behringer states:

> Only when it became normal practice in the business world did it come down in price and become accessible to much wider layers. The arrival of mass tourism in the 1970s brought a rapid increase in numbers of passengers and destinations as well as the capacity of individual aircraft. Energy consumption per passenger rose disproportionately, however, as an aircraft holding three hundred needs as much fuel as tens of thousands of Volkswagen Beetles. (Behringer 2010:178)

World air travel during the period increased from 31 million passengers and 28 billion passenger kilometers in 1950 to 2.022 billion passengers and 3,720 billion passenger kilometers in 2005 (Chafe 2008:71). While international air travel rose 3.8 percent and 3.0 percent in Europe and North America, respectively, between 2012 and 2013, it grew 12.1 percent in the Middle East and 8.1 percent in Latin America over the same period (IATA 2013). Domestic travel increased 4.9 percent world-wide in the period 2012–2013, with a growth rate in China during this period of 11.7 percent. Reportedly, "African airlines international air travel expanded 5.5% in 2013, a solid result but slower than growth in 2012 (7.5%)" (IATA 2013:2). In addition to passengers, airplanes transport many consumer items, including fresh fruit, vegetables, fish, and cut-flowers, many of them from developing countries.

Airplane travel presently represents about 2–3 percent of CO_2 emissions from transportation. Airplanes also emit nitrous oxide and other contrail or exhaust fumes, meaning that a "factor between two and three is normally applied to the CO_2 impact" (Tickell 2008:41). Clark maintains in addition to CO_2 emissions,

> Plane engines also generate a host of other outputs, including nitrous oxide, water vapour and soot. At flying altitudes in the upper troposphere and lower stratosphere, these outputs produce a range of climatic effects. For example, the nitrous oxide at high altitudes causes formation of ozone—a greenhouse gas that warms the local climate—but at same time undermines reactions which destroy methane, thereby removing another greenhouse gas from atmosphere. (Clark 2009:169)

A flight from London to Hong Kong return per passenger reportedly results in 3.4 tons of CO_2e per economy class seat, 4.6 tons on average, and a whopping 13.5 tons for a first class seat (Berners-Lee 2010:135). According to Clark (2009:14) a flight between London and Edinburgh per passenger results in 140 kg of CO_2e in emissions, whereas a one passenger trip in a Ford Mondeo 2.0 results in 120 kg of CO_2e in emissions, a trip in a Toyota Prius with four passengers results in 16 kg CO_2e in emissions per passenger, a trip on an ordinary train results in 15 kg of CO_2e in emissions per passenger, and a trip in a coach results in 18 kg of CO_2e emissions per passenger. While technological innovations have and will probably continue to result in more energy-efficient airplanes, these developments will be offset by the drive on the part of individuals to fly more and more and to encourage more and more people to fly. Fuel efficiency does increase with the size of the airplane, meaning that flights in small aircraft, particularly private jets, are energy-intensive and intensive in terms of greenhouse gas emissions. (Bridger 2013:13).

Although the UK government set a target of reducing CO_2 emissions 80 percent, its 2003 white paper on aviation projected an increase from 200 million passengers moving through UK airports to as many as 400 million by 2020 and 500 million by 2030 (Cahill 2010:5). To facilitate this projected increase, it had planned to expand facilities to be able to handle the increased number of flights.

As in many other areas of a stratified world system, the affluent contribute much more overall to greenhouse gas emissions from flights than working-class people and particularly the poor around the globe. In the case of the UK, the Oxford Transport Studies Unit confirmed this grim reality in its research:

> Although over half of the UK population now travels by air at least once a year—though almost half do not—a very small percentage of people travel many times as often. . . . Cheap air travel may seem to be a great leveller, making long-distance travel available to all; but its most important impact has probably been to allow the richest few percent of the population almost unlimited freedom to pollute as much as they want, barely thinking about the financial impact. (cited in Goodall 2007:183–184)

Given the present predilection for ongoing corporate growth and globalization, what type of future can we foresee for airplane travel and airports as their hubs? John Kasarda has coined the term *aerotropolis* for what he foresees as a world of new cities built around airports that interconnect with numerous other *aerotropoli* around the world, just as seaports and railways served as urban hubs in the past (Kasarda and Lindsay 2011). The Persian Gulf region of the Arabian Peninsula has already evolved into the hub of several aerotropoli, such as Dubai and Abu Dhabi in the United Arab Emirates, which expected to "receive 280 million passengers a year by 2015" (Kasarda and Lindsay 2010:309). Peter Adey (2009) views airplanes and airports as the "signifier of a contemporary mobile world." He maintains that the issue of environmental sustainability has important implications for air travel in the future.

While there has been significant improvement in the energy efficiency of airplanes over the past few decades, in terms of their greenhouse gas emissions, this has been offset by a rise in the number of airplane flights during this period, one more example of the rebound effect or Jevons Paradox. Evans and Schaefer report:

> Passenger demand increases, on average, by 0.07% for every percent reduction in aircraft energy use. This is partly a consequence of the decline in airfares, on average, by 0.24% for every percent reduction in aircraft energy use, which comes with the introduction of low fuel burn technology in a competitive market. (The decline in airfare is smaller than the reduction in fuel burn because fuel burn is only one cost element incurred by the airlines). (Evans and Schaefer 2013: 163)

Ships

Marine transportation has been sometimes posited as a reasonably sustainable sector for moving freight compared with airplanes, trucks, and railroads. George reports:

> Trade carried by sea has grown fourfold since 1970 and is still growing. In 2011, the 360 commercial ports of the United States took in international goods worth $1.73 trillion, or eighty times the value of all U.S. trade in 1960. There are more than one hundred thousand ships at sea carrying all the solids, liquids, and gases that we need to live. (George 2013:3)

Much of this increase has been facilitated by some 6,000 container ships, the largest of which can carry 15,000 boxes. The container ship constitutes a "precondition for global neoliberalism and the rise of China as a world leader in commodity exports" (Eriksen 2016: x).

International shipping was estimated to contribute between 1.6 and 4.1 percent of world CO_2e in 2007 (Psaraftis and Kontovas 2009). The International Maritime Organization (2008) reports that international shipping emitted 84.3 million metric tons of CO_2 and 2.7 percent of global CO_2 emissions in 2007. Including domestic and fishing vessels larger than 100 gross tons, the figure was 1.019 billion metric tons or 3.3 percent of global CO_2 emissions. According to Stanford (2008:262), "Long-distance transportation associated with globalization consumes vast quantities of fossil fuels. Many products now use more energy being transported to far-off markets than they do being produced. Intercontinental ocean shipping . . . is one of the most polluting industries on the planet." Sea shipping constitutes the "circulatory system of the global economy" in which about "90 percent of the world trade products are carried at some point" (Paskal 2010:80). Sea shipping increased between the early 1960s and 2006, going from less than 6 trillion ton-miles to 33 trillion ton-miles. The cost of shipping has decreased by about 80 percent in 25 years due to "containerization, bigger ships, and commuter-assisted resource allocation" (Paskal 2010:81). Williams (2010:226) reports: "Huge numbers of container ships now ply back and forth between continents: polluting the seas and atmosphere, and introducing non-indigenous invasive species." Developed societies have moved much of their heavy and light industry to China to reduce labor costs and then China exports manufactured products back to developed countries. While the shipping industry has become more energy efficient, it constitutes a classical example of the Jevons Paradox in which within a capitalist economy increased production follows suit. Angus reports:

> The bunker fuel used by large cargo ships is the cheapest and dirtiest fuel available: it's as thick as asphalt, made from the waste that's left after all other fuels have refined from crude oil. CO_2 emissions from ships burning bunker have grown 3.7 percent a year since 1990. A large container ship burns 350 tons of fuel a day and emits more CO 2 each year than many coal-fired power stations. (Angus 2016:166–167)

Journeys on some types of passenger ships may result in higher greenhouse gas emissions per passenger than airplanes because of the amount of weight that the former carry, which in the case of some ferries would include motor vehicles, including trucks (Bridger 2013). Cruise ships constitute "floating holiday resorts" (Bridger 2013:15). The *Queen Mary 2* which operates between Southampton and New York requires six days for a crossing with up to 2,600 passengers, consumes 400–1,200 tons of fuel, "making the overall carbon footprint of each ticket substantially more than a seat on plane (Clark 2009:175). Some container ships have room for around twelve passengers providing them with basic amenities and thus may be a more environmentally friendly form of ship travel. While they require a lot of fuel for transporting freight, they are "not as energy-profligate as cruise ships" and are "working boats that will be making the same trips regardless of whether a few passengers come along for the ride" (Clark 2009:175–176).

CONCLUSION

This chapter examines the role of motor vehicle production in the depletion of natural resources and environmental degradation, including air pollution and greenhouse gas emissions, the utilization of valuable farmland in the production of biofuels, and the disposal of motor vehicles. It also briefly discusses the impact of trains, airplanes, and ships on the environment. Motor vehicle companies supported the Coalition for Vehicle Choice in the United States and Global Climate Coalition and Climate Council, international climate denialist lobby bodies. The motor vehicle industry constituted a key partner in the Global Climate Coalition, an industry lobby group that operated out of the Washington, DC (GCC), office of the National Association of Manufacturers. Along with a small number of contrarian scientists, it questioned the findings of climate science, including reports of the Intergovernmental Panel on Climate Change. After British Petroleum and Shell Oil abandoned the GCC in 1999 due to concern that membership would damage their public image, Ford and Daimler-Benz joined various other industries in bailing out, resulting in the collapse of the association (Gelbspan 2004:40). Motor vehicle manufacturers opposed the California state government for devising a plan to reduce greenhouse gas emissions emanating from cars (Beder 2006:17).

As the reality of the ecological crisis and climate change have come to be accepted by large sectors of the public around the world, like other multinational corporations, motor vehicle companies assert that they are concerned about mitigating these phenomena to which they have contributed. US motor vehicle companies left the Global Climate Coalition prior to its demise in 2002 and now all motor vehicle companies support industry organizations such as the World Business Council for Sustainable Development (Mikler 2009:6). Automobiles companies assert that they are "keen to develop 'environmentally-friendly cars' that either 'pollute little,' or are not powered at all by oil, but by 'sustainable fuels,' biofuels, electricity or fuel cells" (Tziovaras 2011:27). They claim to be concerned about "fuel economy and CO_2 emissions reductions, and highlight technology-driven solutions with conventional engines" (Mikler 2009:69). *Mobility 2030* published in 2004 is a report sponsored by twelve corporations, eight of them motor vehicle companies, two of them major oil companies, along with Michelin, Norsk Hydro ASA which serve as suppliers to the motor vehicle industry. The report delineates objectives to reduce air and noise pollution, motor vehicle fatalities and injuries, congestion while improving mobility by promoting the construction of more roads (Beder 2006:16).

In his superb volume on the "fossil economy," Andreas Malm (2016:7) astutely observes that automobile dependence has been fostered by a "vast infrastructure of oil terminals, petroleum refineries, asphalt plants, road networks, gasoline stations—not to speak of the film industry, the lobbying groups, the billboards" which marginalized other forms of transportation. Of course, multinational corporations and their political allies around the world facilitated this dependence. While automobiles may provide some people with temporary relief from heat waves that will more and more plague particularly urban areas which constitute "heat islands" by escaping to the beach or the mountains, the "emissions produced by the cars running to and fro, meanwhile will have their greatest impact on generations not yet born" (Malm 2016:7). The topic of heat island effect has major implications for the impact of climate change on health, an issue that I discuss in the next chapter.

Chapter Five

The Political Ecological Impact of Motor Vehicles on Health

Motor vehicles have had major impacts on not only patterns of consumption, settlement patterns, energy utilization, the natural environment, and, last not least, health, both physical and mental, bearing in mind that they are intertwined. When automobiles first began to appear, various public health reformers viewed them as a panacea to the manure and urine deposited by horses in cities. McCarthy (2007:22) notes: "Some physicians gushed about the benefits of motoring for all sorts of ailments, including baldness and dandruff. One English physician, considering airborne bacteria, even argued that automobile exhaust was an excellent disinfectant."

Conversely, by 1910 some American physicians maintained that automobile exhausts constituted a public health problem (McCarthy 2007:25). Despite much evidence that motor vehicles pose not only significant environmental and health hazards, regulation of motor vehicle emissions did not appear in the United States until the 1960s. None other than Woodrow Wilson during his presidency of Princeton University acknowledged the health hazards of automobiles in February 1906 by indignantly asserting: "I think that of all the menaces of today the worst is the reckless driving of automobiles. In this the rights of the people are set at naught. When a child is run over the automobilist doesn't stop, but runs away" (quoted in McCarthy 2007:12).

In the current era, automobiles have become death machines and transformed roads into "killing fields." Automobile accidents reportedly result in over 1.24 million deaths per year worldwide and an estimated 20 million to 50 million injuries, of which 91 percent occur in low- and middle-income or developing countries where safety standards are not as developed as in high-income or developed countries (World Health Organization 2015). Road traffic accidents are projected to increase from 1.3 million in 2004 to 2.4 million in 2030 due to the anticipated increase in number of motor vehicles worldwide (Lynch and Smith 2005). Unfortunately, motor vehicle accidents are so commonplace that they tend to be taken for granted as part of modernity. As Short and Pinet-Peralta (2010:43) observe, "they are just too mundane, too anonymous, too ordinary to generate much interest."

Woodcock and Aldred highlight the complex linkages between car-based transportation and public health:

> The burning of fossil fuel, oil, produces the kinetic energy that kills 1.2 million people and injures 50 million on the roads each year. Urban pollution for motor vehicles is responsible for hundreds of thousands of deaths per year, while noise pollution is implicated in cardiovascular disease, and inhibits cognitive development. The obesity epidemic is linked to the shift from human powered to fossil fuel power transportation, which contributes to an imbalance between energy expenditure and energy consumption. Major roads can sever communities by barring local access, communication and social integration, with implications for social capital. These harms and benefits are unequally distributed. (Woodcock and Aldred 2008:3)

PHYSICAL HEALTH

Motor vehicles are a major source of deaths and injuries around the world. In the current era, automobile accidents reportedly result in more than 1.2 million deaths per year worldwide and up to 50 million injuries, the vast majority of which occur in low- and middle-income or developing countries where safety standards are not as developed as in high-income or developed countries (World Health Organization 2015). The World Health Organization also reports two other gripping statistics:

- "Globally, road traffic crashes are a leading cause of death among young people, and the main cause of death among those aged 15–29 years" (WHO 2015:x).
- "Data suggest that road traffic deaths and injuries in low- and middle-income countries are estimated to cause economic losses of up 5% of GDP. Globally an estimated 3% of GDP is lost to road traffic deaths and injuries" (WHO 2015: xi).

Road traffic accidents are projected to increase from 1.3 million in 2004 to 2.4 million in 2030 due to the anticipated increase in number of motor vehicles world-wide (Lynch and Smith 2005). Urry (2007:207) boldly asserts that the "freedom to drive is the freedom to die." In a similar vein, Woodcock and Aldred (2008:5) propose that the term *motorized violence* is a more appropriate designation for what generally are termed *road accidents* for a phenomenon that has become "socially naturalised" and for which "justice systems in many countries impose minimal penalties upon drivers causing death or serious injury." They also assert that the "threat of violence is embodied through obesity, diabetes, and cardiovascular disease," ailments that in large part are a consequence of diminished active travel that contributes to the obesity epidemic (Woodcock and Aldred 2008:6). In other words, heavy reliance on cars as a mode of transportation contributes to less exercise through walking and cycling and thus to obesity and various diseases.

Urry (2007:118) reports that "[c]rashes are a normal and predictable outcome of the car system although they are typically referred to as 'accidents,' aberrations rather than 'normal' features of what Beckmann [2001] terms the 'auto-risk society.'" Forman, Watchko, and Sequi-Gomez (2011) refer to deaths and wounds from automobile collisions as an "overlooked epidemic" deserving of medical anthropological study. Indeed, their very hazardous nature began to be apparent very early in their history. Brian Ladd reports:

> In 1906–07, over 90 percent of the two thousand cars registered in Berlin were involved in accidents. Too often the victims were pedestrians. Most were poor and anonymous, but not all: in 1914 two members of the German parliament were run down on a Berlin street. Already a member of the Prussian parliament had complained that "there is hardly a person in Berlin who has not been on the verge of being run over." (Ladd 2008:73)

In the United States, as late as the 1920s there was a tendency for juries arbitrating motor vehicle accidents to rule in favor of injured pedestrians. According to Garrison and Levinson (2014:198), partly "this was a class issue: the average motorist was a wealthier pedestrian, or jury member." Eventually local safety councils urged motorists to drive more safely and subtly shifted the blame for accidents from the motorist to the pedestrian.

In contrast to accidents incurred by automobiles, those incurred by motorcycles are generally much higher. As Packer (2008:113) astutely observes, the "motorcycle could be considered that which makes the automobile accident rate acceptable." Particularly in the United States but also in Australia, motorcyclists have come to be viewed particularly since World War II as reckless, rebellious, dangerous, and beyond the pale of the law, although efforts have been made in motorcycle advertisements to depict at least some motorcyclists as nice people (Packer 2008:122–123).

Historically automobile fatality rates have varied over time. For example, they dropped during World War II in the United States after President Roosevelt suspended domestic automobile production in February 1942 to shift the infrastructure of the automobile toward manufacturing military hardware. Consequently, as opposed to 38,142 auto fatalities in 1941, they dropped to 22,727 auto fatalities two years later, a 40 percent decline (Blanke 2007:194). Even in the Soviet Union where there was a paucity of motor vehicles of various kinds compared to developed capitalist countries they became a rising cause of accidents during the late 1920s and 1930s, although some of the evidence in this area is spotty. Whereas in 1927, there were 2,292 accidents due to all forms of motor vehicles, this had increased to 7,703 accidents in 1936 (Siegelbaum 2008:197).

Automobile manufacturers have resisted legislation and public pressure to build safer cars. O'Connell and Myers (1966) assert that they have attempted to shift responsibility for the increasing of accidents to driving behavior and road conditions. One of the most profound indicators of the fatalities incurred in automobile accidents are roadside memorial markers that have appeared in the United States and possibly other countries. (Blanke 2007:184). Chakraborty conducted an epidemiological study of the distribution of cancer and respiratory risks from exposure to motor vehicular pollutants in the Tampa Bay Metropolitan Statistical Area that found African Americans and Hispanic Americans were at a greater risk than were European Americans in the area, even after controlling for socioeconomic status because the former generally reside near major thoroughfares. Furthermore, "densely populated neighborhoods in the more urbanized parts of the MSA are more likely to experience higher traffic volumes and are thus exposed to greater health risks from vehicular emissions" (Chakraborty 2009:690).

Motor Vehicle Fatalities

Fallon and O'Neill (2005) claim that the world's first automobile accident, which involved a steam carriage, occurred in the Irish Midlands in 1869 when Mary Ward fell out of the vehicle and was run over. Another early automobile fatality in Britain occurred in July 1899 when Edwin Sewell and his passenger Major Richie died due to the rear wheel of their vehicle collapsing, throwing them out (Mosey 2000:32). Similar instances prompted distrust in automobiles, including on the part of Queen Victoria. Reportedly the first automobile fatality in the United States occurred in when Henry Bliss stepped off a trolley car in New York and was hit on September 13, 1899, by an automobile, dying of injuries the following day (Black 2010:55). In 1900 there were 36 automobile fatalities in the United States and by 1925, the annual fatality figure from motor vehicle accidents hit 20,771 (Black 2010:55). In New York City, fatalities due to automobiles came to exceed those due to horse-drawn wagons and street cars. Whereas in 1910, 148 fatalities resulted from streetcars and 201 as a result of horse-drawn wagons, two years later, "automobiles killed 221 people, exceeding both wagons (176) and streetcars (135)," the great majority of whom were pedestrians (Wells 2013:88). Whereas some 260,000 Britons were killed as a result of World War II between 1939 and the 1945, by 1984 more Britons, namely 287,000, had died in road accidents over a 75-year period (Hamer 1987:141).

Freund and Martin present the following sobering statistics on auto accidents resulting in deaths:

> While the death rate due to auto accidents in the United States is by no means the highest among industrialized countries, some 43,000 to 53,000 Americans die each year in such accidents, producing a death rate of 26 deaths per 100,000 population. Worldwide, some 200,000 people died in traffic accidents in 1985. There are approximately 4 to 5 million injuries related to motor vehicles in the United States. Of these, 500,000 people require hospitalization. . . . Auto accidents are a leading cause of death for young people between the ages of five and twenty-four; young males between the ages of fourteen and twenty-four are at highest risk. Per passenger mile, cars are more dangerous than trains, buses, or planes. (Freund and Martin 1993:59)

Over the course of the twentieth century some 3 million motor vehicle-related fatalities occurred in the United States alone (Black 2010:55). Leonard Evans, a General Motors Research researcher, argues that in comparison to the UK, Canada, and Australia, US automobile safety policy emphasizes survival from motor vehicle accidents rather than elimination. Black (2010:56–57) reports: "In 2007 estimates suggest that about 2.49 million people are injured annually in motor vehicle-related accidents in the United States. Of this number, 2.24 million were occupants of motor vehicles, 103,000 were motorcyclists, 70,000 were pedestrians, and 43,000 were bike riders. Based on 2003 data, motorcyclists have the largest number of injuries per hundred million passenger miles, with 555. Automobiles are second, with 69, and light trucks (including SUVs) are third with 51." Globally it has been estimated that some 30 million people died in automobile accidents over the course of the twentieth century (Alvord 2000:114–115). Although the Soviet Union never had the number of automobiles on a per capita basis that advanced capitalist societies had over the course of the twentieth century, the gradual increase in automobiles in the former resulted in a significant increase in the number of road fatalities, but rates varied for the different republics (Siegelbaum 2008:257). In 1990 420,000 people were killed and some nine million were injured around the world due to motor vehicle accidents, and between 1960 and 1994 approximately five million died as a result of motor vehicle accidents. Road traffic accidents are reportedly the leading cause of death worldwide among males between ages 15 and 44 (Crawford 2002:70).

In their survey of cities in various parts of the world, Newman and Kenworthy (1999:118) report that traffic deaths in 1990 in US cities, "despite their highly developed road systems, strictly regulated traffic, and a population generally well educated in traffic safety issues," were the highest at 14.6 per 100,000, compared to 12 per 100,000 in Australian cities, 6.5 in Toronto (Canada), 8.8 in European cities, 6.6 in "wealthy" Asian cities, and 13.7 in "developing" Asian cities. Furthermore, "Amsterdam, at 5.7, and Copenhagen, at 7.5 deaths per 100,000, have among the lowest rates in Europe and have among the highest rates of bicycle usage" (Newman and Kenworthy 1999:119).

Although SUVs are often perceived as safe due to their large size, particularly their early models (such as the Bronco II), they are prone to rollovers that kill and injure occupants at an alarming rate. Bradsher (2002:163) reports that SUV rollovers killed some 120,000 people in the United States during the 1990s. Despite design improvements that have made SUVs safer for occupants, their large size continues to pose a hazard to other drivers and pedestrians.

Recent figures of road traffic deaths per 100,000 population are 24.1 for what the World Bank terms low-income countries, 18.4 for middle-income countries, 9.2 for high-income countries, and 17.4 world-wide (World Health Organization 2015:5). In terms of WHO regions, the road traffic fatality rates per 100,000 population are 26.6 for the African Region, 19.9 for the Eastern Mediterranean Region, 17.3 for the Western Pacific Region, 17.0 for the South-East Region, 15.9 for the Region of the Americas, 9.3 for the European Region, and 17.4 for the World (World Health Organization 2015:6). The WHO reports that nearly half of all fatalities on the world's roads occur among motorcyclists, cyclists, and pedestrians, although the rate in this varies from region to region, with the highest figure being for the African Region (World Health Organization 2015:8). In the case of the United States, David Blanke (2007:3) observes: "The most painful and direct cost of mass automobility is the toll in human lives. Between 1899 and 2003, nearly 3.2 million Americans died due to automobile accidents. In 2001, motor vehicles were the single leading cause of death for Americans between ages sixteen and thirty-four, and the eighth leading cause of death for all ages." The bicycle-related mortality increased 99 percent in Shanghai from 1992 to 2004 because that cyclists had to compete with motor vehicles (Calthorpe 2016:96).

In the case of cities in developing countries, motor vehicle-related fatalities vary widely, depending on various local conditions. Table 5.1 depicts motor vehicle-related fatality rates for three cities, namely Hanoi in Vietnam, Pune in India, and Xi'an in China.

Table 5.1. Motor Vehicle-Related Fatality Rates in Hanoi, Pune, and Xi'an

Indicators	Hanoi	Pune	Xi'an
Fatalities/year	503	282	545
Fatalities/million people	152	76	76
Fatalities/1000 vehicles	0.46	21	1.05
Fatalities/million passenger kilometers	22	6.3	7.06

Source: Partnership for Sustainable Urban Transport in Asia (n.d.: 32).

Hanoi (population 3.1 million in 2003) has by far the greatest number of motorcycles and mopeds of the three cities. Pune (population 2.7 million in 2003) is the headquarters of Bajaj, one of the world's leading manufacturers of two- and three-wheel motor vehicles, exhibits a "high density of vehicles on the roads everywhere and erratic driving make driving, walking, or cycling hazardous" (CAI-Asia Program) (Partnership for Sustainable Urban Transport in Asia, n.d.: 33). Xi'an (population 5.1 million in 2003) has a world-renowned inner-city wall and has a higher density of both automobiles and buses and lower density of motorcycles and mopeds than both Hanoi and Pune, translating into the highest rate of fatalities/million passenger-kilometers of the three cities.

Blanke provides interesting historical evidence comparing the fatalities rates for selected infectious diseases with those emanating from motor vehicle accidents between 1910 and 1940 in the United States, as delineated in table 5.2 below.

While public health measures and medical treatment have during the early twentieth century contributed to declining mortality rates from infectious diseases, the rising of utilization of motor vehicles during the same period contributed to a rising mortality rate from their use.

People who study traffic safety term "vulnerable road users," namely pedestrians, cyclists, and vehicles in which drivers are less likely to use child

Table 5.2. Fatality Rate for Selected Infectious Diseases and Motor Vehicle-Related Cause, 1910–1940, per 100,000 Population in the US*

Year	Syphilis	Tuberculosis	Whooping Cough	Typhoid	Motor Vehicle Accidents
1910	13.5	153.8	11.6	22.5	1.8
1920	16.5	113.1	12.5	7.6	10.3
1930	15.7	71.1	4.8	4.8	26.7
1940	14.4	45.9	2.2	1.1	26.2

*Adapted from Blanke (2007:99)

safety seats and seat belts as more likely to be either killed or injured in a motor vehicle accident. As Lutz (2014:239) states, "many car crashes are occupational hazards associated with lower-paid work in agriculture, forestry, ground transportation and delivery services, mining, and construction."

While working-class people are probably more apt to die as the result of automobile accidents, Mitika Brottman (2001:xv) asserts that the "car crash is the archetypal means of *celebrity death*" in that from "motorcade and gold-plated Rolls to tour bus and stretch limousine, celebrities are particularly fond of their automobiles. Nothing makes a better tabloid headline than the celebrity car crash, especially if it involves drugs, alcohol, excessive speed, violence, or passengers who shouldn't have been there. Nothing ends a tale of beauty, wealth, and potential better than blood on the tracks." The death of Princess Diana, which occurred riding with a drunk driver at a reported speed of 196 kilometers per hour became news around the world and the subject of close-up photographs snapped by paparazzi (Duffy 2009:2).

R. J. Smeed (1949), a British statistician and road-safety expert, based upon research done in 20 developed countries, concluded that the number of motor fatalities on the road rises to a certain point, but then the fatality rates begin to drop and even the absolute number of fatalities drops. This is because, with rising fatalities, governments begin passing regulations that promote safety, such as the requirement to wear seat belts and restrictions on the amount of alcohol that a driver may consume, which may make drivers more cautious in their road behaviour (Vanderbilt 2008:231). Siem Oppe (1991:410) maintains that Smeed formulated his generalization at a time when automobile traffic was still on the rise, but that in some developed countries, particularly Japan, the "development of traffic volume is already reaching its point of saturation." Furthermore, it is not clear whether Smeed's generalization applies to developing societies, many of which have just begun to increase in traffic volume as more and more particularly middle-class people adopt private cars. Urry (2007:207) boldly asserts that the "freedom to drive is the freedom to die." In a similar vein, Woodcock and Aldred (2008:5) assert that the "threat of violence is embodied through obesity, diabetes, and cardiovascular disease," ailments that in large part are a consequence of diminished active travel that contributes to the obesity epidemic (Woodcock and Aldred 2008:6). In other words, heavy reliance on cars as a mode of transportation contributes to less exercise through walking and cycling and thus to obesity and various diseases. An epidemiological study by Sovacool (2010) revealed that in the US in 2008, 36,710 deaths resulted from automobile fatalities while another 65,638 deaths resulted from particulate-matter pollution, much of it due to motor vehicle exhaust fumes.

Table 5.3. Estimates of Road Fatalities in Lower- and Higher-Income Countries

	1990	2000	2010	Change by decade	
				1990–2000	2000–2010
Fatalities (1000s)					
Lower-income	419	613	862	46%	41%
Higher-income	123	110	95	–11%	–14%
Total	542	723	957	33%	32%
Fatalities per 100,000					
Lower-income	9.6	12.1	14.9	26%	23%
Higher-income	13.6	11.3	9.3	–17%	–17%
Total	10.3	12.0	14.1	16%	18%

Source: Adapted from Gilbert and Perl (2010:213).

Table 5.3 depicts estimates of road fatalities in lower-income and higher-income countries.

Table 5.4 below compares road fatalities and homicide rates for selected regions and countries around the world.

A simple comparison of the United States and Canada in terms of mortality rates of both homicide and road crashes reveals that the United States on both counts is a more violent society than its neighbor to the north which has gun control laws and apparently somewhat safer drivers and/or driving regulations. Although the United States with its minimal gun control laws is commonly viewed as a violent society, more of this violence is perpetrated by motor vehicles than weapons. Public awareness of the dangers of automobiles reached new heights in the United States with the publication of Ralph Nader's (1965) book *Unsafe at Any Speed*. Although there had been

Table 5.4. Comparison of Road Fatalities and Homicide Rates*

	Mortality rate per 100,000 persons		
	Homicide	Road crash	Ratio
France	0.7	12.1	17.3
Japan	0.6	7.4	12.3
China	1.8	19.0	10.6
Canada	1.4	9.3	6.6
USA	6.9	15.0	2.2
Sweden	1.2	2.6	2.2
Mexico	15.9	11.8	0.7
Russia	21.6	9.7	0.4
Colombia	61.6	24.2	0.4

*Adapted from Gilbert and Perl (2010:215).

efforts to reduce motor vehicle accidents with the installation of seat belts and other safety devices and, at least until 1995, the lowering of speed limits, such measures tend to require air bags in cars. The victim of an automobile accident is, in reality, the victim of a transportation system largely framed around profit-making.

Speed constitutes a major contributing factor to motor vehicle deaths and accidents. The implementation of a short-lived 55 miles per hour speed limit in the United States reduced highway fatalities 20 percent (Wolf 1996:172). Germany is famous or infamous, depending upon one's perspective, for the lack of a speed limit on its autobahns, a policy that was first established during the Nazi era. Whereas the German Democratic Republic (East Germany) had a speed limit on its autobahns, the unification of 1990 resulted in the eradication of this restriction and a doubling of the number of highway fatalities (Wolf 1996:172).

Mortality rates resulting from motor vehicle accidents in developing societies tend to exceed those in developed societies. As Vanderbilt (2008:231–232) observes, "In China, one sees things . . . like bicycles traveling on restricted highways, scooter drivers carrying several children without helmets, and drivers stopping on the highway to urinate—but presumably, a number of years down the road, these things will largely be only memories." Hopefully so, but driver safety regulations will not eradicate the pollution and greenhouse gases emitted by motor vehicles in a country with a rapidly growing number of motor vehicles. While China has 2 percent of the world's cars, it has 15 percent of the world's traffic fatalities (Kingsley and Urry 2009).

In some developing countries, motor vehicle fatalities result not so much from automobiles but motorcycles. In Vietnam, for example, "[T]here were 4 million of these vehicles registered in 1995 but 14 million by 2004. In the same period road traffic fatalities increased five times. In Hanoi, for example, roads are very narrow: motorcycles and pedestrians often share the same public space; speed limits are often broken; stop signs and pedestrian crossings are rarely adhered to. . . . Motorcycles are often overloaded with multiple family members, and child passengers rarely use crash helmets" (Short and Pinet-Peralta 2010:48).

Injuries from Motor Vehicle Accidents

The motor vehicle accident rate appears to be the highest in developing countries, which have more recently acquired large numbers of cars and other motor vehicles compared with developed countries. Cahill (2010:35) argues that "we see newly emerging consumer societies such as China and India promoting the car at the expense of the bicycle." Ironically, many cyclists probably find themselves so endangered by the growing onslaught of cars and even buses, jitneys, motorcycles, and motor scooters in developing countries that they feel forced to capitulate to motorized vehicular transport.

Wang and Dalal (2012) conducted an epidemiological study which examined road traffic injuries in the city of Shanghai. Of 1,205 individuals who were involved in a total of 568 traffic accidents, 2 percent died, 43 percent sustained mild injuries, 7 percent received severe injuries, and 48 suffered no injuries. Wang and Dalal (2012:79) report: "Most victims were passengers, motorcyclists and pedestrians. Road accidents mainly occurred at night between 19–20 pm [sic – should be 7–8 pm] being peak hours, and most deaths occurred at 3–5 am (Wang and Dalal 2012:79)." In developed countries and probably in developing countries, motor vehicle fatalities on rural roads are considerably higher relative to population than in urban areas. In the case of the United States, "rural areas, as defined by traffic safety analysts, contain 23 percent of the national population but witness 56 percent of fatal crashes" (Meyer 2013:94).

Anthropologist Judith Barker (1999) presents a revealing case study of road traffic crashes on Niue Island in the South Pacific. She asserts: "Most Pacific (e.g., Solomon Islands, Papua New Guinea, Western Samoa) exhibit road traffic crash (RTC) fatality rates consonant with low motorization but high with respect to total number of vehicles in country: that is, the death rate from RTCs is high with respect to the total number of vehicles in the country. . . . By these measures, Niue is a Pacific nation with a high level of motorization and considerable societal costs from RTC" (Barker 1999:214). Barker gathered her data over twelve months in 1982–1983 when Niue had a population of some 2500 people, with most Niueans having opted to migrate to New Zealand (where they hold citizenship). In comparison to most Pacific states, Niue had a relatively high material standard of living because more than 80 percent of the population was employed in the service sector or in government or received remittances from expatriate relatives in New Zealand (but less so today as 90 percent of Niueans reside in New Zealand). These factors allowed many Niuens to own motor vehicles of one sort or other. Many young men and women drove motorcycles which were hazardous to drive because the major roads consisted of only two lanes and the road surface often was rough. Young men particularly mixed motorcycle riding with drinking and competition for young women. Barker estimates that nearly half of all RTC on Niue in 1982–1983 were alcohol-related. She observes: "In 1982, some 11.2% of all adult male hospital admissions were the result of RTCs, matching some of the highest RTC admission rates in the world" (Barker 1999:228). Thus, while we generally associate traffic accidents and fatalities with urban life, the Niuen case indicates that even in a supposedly idyllic remote setting in the South Pacific motor vehicles have had adverse impacts.

OTHER MOTOR VEHICLE-INDUCED HEALTH PROBLEMS

Other motor vehicle-induced health problems include respiratory and musculoskeletal ailments, stress, lowered cognitive development in children, and hearing loss due to noise pollution from traffic. As early at 1910 some physicians asserted that automobile emissions posed a public health threat but serious consideration of the problem had to wait several decades to mount (McCarthy 2007:25). In the 1920s tetraethyl lead was added to petroleum in order to reduce knocking, and increase octane rating as well as power fuel economy. However, when combusted lead became an air pollutant and could lead to higher concentrations in the blood stream. This was not only bad for general health but contributed to lower cognitive development in children (Garrison and Levinson 2014:414). Ironically, although lead in gasoline was not banned in the United States until the 1970s, the Soviet Union did so in cities in 1967.

Research Atlanta based at Georgia State University has found that "asthma-related visits to the pediatric emergency clinic at Grady Memorial Hospital go up by a third during Atlanta's high ozone days" (Doyle 2000:3). The Centers for Disease Control and Prevention (CDC) estimated that the number of asthmatics in the United States increased from 6.8 million in 1980 to 17.3 million in 1998, a fact in large part related to an increase in the amount of motor vehicle traffic in urban areas (Doyle 2000:5). Due to legislation, emissions from motor vehicles have declined overall in the United States and other developed countries.

Lutz (2014:239) maintains that "[p]eople residing inside major road corridors are at higher risk of heart and lung disease, childhood cancers, brain cancer, and leukemia, and the class composition of the neighborhoods in such corridors tends strongly to the poor." Driving causes increased level of stress hormones, blood sugar, and cholesterol in the body (Freund and Martin 1993:33). Driving an automobile or even riding in one often induces stress due to anticipating or avoiding danger, waiting for traffic lights to change, and being in congested traffic. Bus drivers and fare collectors on buses also experience stress due to the need to be constantly alert.

Motor vehicle driving, particularly under congested conditions, also induces stress and contributes to medical complications such as lumbar disk herniation, or motorist's spine, and contributes to a sedentary lifestyle. Truck drivers in particularly suffer a high rate of back injuries. City bus drivers around the world experience a great deal of stress, particularly during peak hours. Reportedly, "[m]edical ailments send more than half of them into early retirement" (Vanderbilt 2008:141). Truck and bus drivers often develop musculoskeletal ailments due to sitting for long stretches of time and from the vibrations caused by their respective vehicles (Freund and Martin 1993:34). I first became aware of this while on a hitchhiking expedition across the plains of western Nebraska. A truck driver who picked me up told me that few of his long-term, long-distance compatriots reach the age of 65, in part due to the fact that the constant vibration of the truck severely damages their internal organs.

Ordinary automobile drivers may also incur musculoskeletal problems, although it is questionable how much of these problems are related to sitting in an automobile or elsewhere, such as at home, an office, theater, classroom, or lecture hall, especially given that so many chairs and seats are poorly designed from an ergonomic perspective. At any rate, Pietri et al. (1992) conducted a study in which they concluded that driving ten hours a week or more induced seat, and psychosomatic factors could induce the first occurrence of low-back pain." Kelsey and her colleagues in a series of studies (1975, 1984, 1990) founded a relationship between number of hours driven per week and elevated risk for acute herniated lumbar intervertebral disk.

Traffic noise reportedly contributes to both physical and mental health problems, including disrupted sleep patterns, hearing damage, anxiety, and even depression (Rylander 1992; Mugyeni and Engler 2011). Bongardt et al. (2013:38) report that "[n]oise and vibration from transport contributes to sleep disturbance, which in turn can lead to increased blood pressure and heart attacks." It is important to point out that other modes of transportation, such as subway trains (particularly when they run above ground in places such as New York City), buses, and trucks constitute sources of noise pollution in urban areas. Wright reports (1992:66), "Most transportation noises are at much lower levels for those on board or near vehicles: passenger cars, 65–85 dB(a); buses, 80–86 dB(A0; trains, 75–95 dB(A); trucks, 80–90 dB (A). Most correctly outfitted motorcycles operate at 78–88 dB(A), but some reach unacceptable levels of about 104 dB(A). At the start of motorcycle races, levels of 115 dB(A) or more are reached. Motorcycles whose mufflers have been tampered with probably come near that high level also."

Trucks generally reach their highest noise levels when cruising on freeways. It is difficult to say to what degree motor vehicle-related noise pollution contributes to long-term hearing loss, but probably more so than is generally assumed to be the case since motor vehicle-related noises have become normalized or seen as part of living in a modern society. Elderly people in various traditional communities retain good hearing for the duration of their lives. According to Wright (1992:66–67), "[w]orkers in vehicle factories, drivers of exceptionally noisy trucks and buses, airport ground crews, paramedics in ambulances, and highway patrol personnel with long hours at the roadside are probably the only persons at high risk of suffering significant or permanent hearing handicaps as a result of traffic noise alone." Even in a seemingly tranquil country like Switzerland, 1.68 million people are subjected to road noise above the daytime threshold limit for residential areas set at 60 dB(A) (Umweltstatistik Schweiz 2009).

Peter Freund and George Martin (2008) explore the historical and structural connections between automobile-centered transport and fast food franchising. The "fast foods" complex began to emerge after World War I in the United States but exploded after World War II, particularly with the development of the mass production of food and the interstate highway system. Fast foods are a major contributor to obesity and its associated health complications. At the same time, motor vehicles have discouraged more active forms of mobility, particularly walking and cycling. Freund and Martin (2008:314–315) also comment upon the ecological footprint of the fast cars/fast foods complex in that cars account for about one-third of the US total energy use and that fast food is transport energy intensive: "Fast food consumption is predicated on the transport of food from production sites to distribution sites and then on to retail outlets, as well as on individual patrons driving their cars. . . . In addition, the industrial agriculture used in the supply chains of fast food corporations is resource and energy intensive. Finally, there are additions to the fast food ecological footprint, such as its heavy use of packaging material."

While the role of motor vehicles in transmitting infectious disease remains a thinly studied topic, Gewald, Luning, and van Walraven (2009) maintain that the advent of the motor vehicle has been a mixed blessing for Africans. Whereas on the one hand it has improved access to inoculation campaigns, primary health care, hospitals, and medical extension work, on the other hand, "these vehicles have, at the same time, become the main vectors for the spread of disease and the speed of transfer of viruses from forest enclaves to cities, and vice versa, along roads transecting the continent has increased markedly. The most notable has, of course, been the rapid and devastating spread of HIV/AIDS" (Gewald, Luning, and van Walraven 2009:8).

Finally, it is important to note that while the motor vehicle production do not generally result in fatalities per se, it may result in psychosocial and musculoskeletal disorders, which Chan et al. (2014) found to be the case in a survey of 1,110 automobile workers at two assembly plants in seven Chinese cities. According to Chan et al. (2014:517), "[r]espondents who think that they were under psychosocial stress report two times more MSDs than respondents who say they do not experience this pressure: 3.86 vs. 1.95." Unfortunately, in China MSD is not included in the list of 115 occupational ailments (Chan et al. 2014:523).

THE POLITICAL ECOLOGICAL IMPACT OF MOTOR VEHICLES ON MENTAL HEALTH

As discussed briefly in chapter 3, motor vehicles have a profound impact upon sociability and psychic well-being, both of which are intricately interrelated. Over the course of the twentieth century, depending in part upon the country and the city or specific geographical locality, such as small rural town and its hinterland, the culture of automobility inculcated in people what various scholars have term "automobile dependence," not only at the physical level of being able to get around but at the emotional level. The French transportation scholar Gabriel Dupuy sketches a "model of structural dependence that lends itself to psychological understanding as well" which entails steps such as obtaining one's driver's license, purchasing one's first car, and driving on roads which draws the individual into a "convenient, familiar, and popular system" (Ladd 2008:163). In turn, psychological dependence on automobiles serves to reinforce structural barriers to public transportation, whether in the form of the deterioration of existing transit systems or reluctance of the powers-that-be, such as politicians and urban planners, to take steps to improve public transit. What this means particularly in countries such as the United States and Australia is that many youth have had few experiences in getting around either by public transit, cycling, or walking, or sometimes all three modes of mobility.

The Automobile as a Source of Sociability

Automobiles came to play an important role in courtship and mating practices in the United States beginning in the late 1920 (Flink 1988:160). Much of youth culture, particularly in developed societies such as the United States, Canada, the UK, other European countries, and Australia, revolves around automobiles given the mobility that they afford young people who are trying to connect with their counterparts in scattered locations not so easy to access by public transportation, particularly at the spur of the moment (Carrabine and Longhurst 2002). Automobiles impacted on women's role as more and more women came to drive as well. In the United States, until the 1920s automobile touring had been severely restricted because of the poor quality of roads in popular tourist destinations, such as national parks (Flink 1988:169–171) However, automobile touring began to take off as a result of the construction of better highways in the 1930s, a development that contributed to the appearance of motor courts in states such as California, Florida, and Texas (Flink 1988:185). During the 1950s and 1960s, drive-in theaters particularly were popular with American teenagers and were referred to as "passion pits," places where undoubtedly an uncountable number of children were conceived outside of wedlock. However, drive-in movie theater operators attempted to portray them as a venue for a wholesome family outing. Foster (2003:86): "Since many theaters charged a fixed price per vehicle, they traded on the idea that tired and overworked parents could take the whole family to double features at bargain prices. Equally important, they would be spared the hassles and expense of locating and hiring baby-sitters. . . . Although industry officials aggressively promoted sales of food and drinks at concession stands, families could further cut corners by bringing their own refreshments—in contrast to management policies in most walk-in neighbourhood theatres." For city-dwellers seeking relief from the hectic pace of urban life, the car and even the camper or recreational vehicle offer an escape to imagined idyllic locations. In reality, most trailer parks in North America, Europe, and Australia are flooded with urbanites who park close to each other while they watch TV, drink, and cook over a camp stove brought from the city, or perhaps simply cook inside their camper vehicles.

Low-income people, who already suffer from the injuries of class both materially and psychically, often find themselves further marginalized without access to adequate transportation in cities where an increasing number of jobs are located in outer suburb areas. Kay (1997:38) maintains: "The car culture has thus become an engine of inequality, raising high the barriers of race and class. Transportation that is difficult at best, non-existent at worse, darkens their lives [of low-income people] in myriad ways and adds to the financial and social inequality they suffer."

Automobile transportation discourages social interaction which at least sometimes occurs in the process of taking public transit and leads to social isolation in that most motorists, especially in advanced capitalist societies, drive alone. As Wolf (1996:192) so aptly observes, "The car society reproduces an elementary phenomenon of the capitalist mode of production: the depersonalization and reification of human relationships. The alienation of the producers (wage earners) in the labour process is reproduced in the organisation of transport." Young people in many countries, particularly developed ones, engage in cruising as a form of sociability, particularly on Friday and Saturday nights. This is no less true in the relatively isolated capital city of Reykjavik on the small and relatively sparsely populated island-nation of Iceland. Like elsewhere, in Reykjavik, cruising serves as a means for interacting with peers and taking one's first steps into the culture automobility. Many of Collin-Lange's (2013:420–421) young informants "explained the importance of good company during the cruise, but also detailed the numerous interactions that they have with other cars and/or people that they meet for example at the ice cream stores."

With the decline of public transportation, especially in the United States, mothers especially serve as chauffeurs for their children as they transport them to various activities in sprawling suburbs. In fact, in less than ten years after 1983, women's automobile travel reportedly quadrupled in the United States (Kay 1997:22). Generations of children in developed societies have internalized car dependence and have been prevented from developing social connections with fellow students who walk to school, often in groups or at least in pairs. In contrast to 1969, when slightly over 40 percent of American children walked or cycled to school, by 2001 only about 15 percent of students were walking to school, a likely factor in increased childhood obesity (US Department of Transportation 2008).

While particularly in developed countries, most motorists drive alone, the presence of passengers in the automobile transforms the journey into a form of sociability. Laurier et al. (2008:2), in conjunction with the Habitable Cars Project, conducted a study in the UK on the "collective private transport" experience in which a "number of people—be they friends, families, acquaintances or colleagues—find themselves sharing a vehicle, more or less, informally" in which they identified the following seven themes about multiple passengers driving in a private motor vehicle:

- Driving constitutes a "socially ordered activity" in which participants engage in specific tasks
- The traditional dinner- or breakfast-time conversation has been shifted into the space of the automobile
- Outside observers, either other drivers or pedestrians, may observe certain elements of sociality in the multiple passenger private vehicle
- The automobile constitutes an important site for "parents to learn about and dialogue with their children, and for children to learn from their parents"
- Commuters sharing a private vehicle take on certain "expectations, obligations towards and values for one another"
- Highly serious conversations about life, love, and death can occur in the confined space of the private motor vehicle
- The car provides the opportunity for collective events such as the proverbial Sunday drive or other outings (Laurier et al. 2008:20)

Ironically, at the same time, low-income people living in inner cities in the United States often find themselves with inadequate access to public transportation to jobs, health clinics, and hospitals, all of which are increasingly located in suburban areas. As previously referenced, Kay (1997:38) aptly observes, "[T]he car culture has thus become an engine of inequity, raising high the barriers of race and class. Transportation that is difficult at best, nonexistent at worst, darkens their lives their lives in myriad ways and adds to the financial and social inequality they suffer." Elderly people with impaired vision and reflexes often continue to drive because the automobile is necessary to maintaining social connections with friends and family scattered about urban areas.

In Luigi Tomba's ethnography of neighborhoods, referred to as Hopetown, consisting largely of rising middle-class people in various Chinese cities, he observed repeatedly that

> buying a car is a high priority for Hopetown's residents, and the growth in the number of vehicles often outpaces the construction of parking facilities. While car sales had grown dramatically in 2002 (133,000 vehicles were sold in Beijing in the first six months of that year), cars became an even hotter consumption item in 2003, with the SARS outbreak convincing many to speed up the purchase of a private car to prevent risks from public transport and taxis. The number of cars circulating on the congested roads of the Chinese capital had already reached the 2 million mark and would surpass 5 million by 2012. (Tomba 2014:112)

In Hopetown 2, car towns protest the high costs of underground parking (an average of 280 yuan or US$33 per month). Furthermore, despite the availability of parking spots in new facilities, many car owners opted to park their cars in public spaces, including gardens, leading to conflicts between the garden-loving residents and the car owners (Tomba 2014:135).

Psychic Well-Being

Driving, along with other forms of mobility, including airplane flights, trains and buses, and even cycling and walking, depending on the circumstances can be stressful, although ironically, each of these forms of mobility can also be enjoyable and relaxing. Commuting particularly, whether by automobile, bus, train, tram, and even airplane, can be highly stressful (Koslowsky, Kluger, and Reich 1995). All forms of transport mobility entail some degree of stress, albeit some more than others. Traffic congestion, crowding on trains, trams, and buses, and standing in line to check in or to go through the security system for airplane flights all can be stressful and frustrating, whether it is for a commute to and from work or even a day excursion for pleasure or a holiday. Much of these pressures are associated with the pace of life in modern capitalist societies where time is of the essence in keeping the economy functioning, especially while commuting to and from the workplace, be it a factory, an office, a café or restaurant if one is a server, or a school if one is a teacher, and so forth. Koslowsky, Kluger, and Reich (1995:38) identify four "environmental stressors" and examples of each embodied in commuting:

- Noise
  - subway entering platform, horn honking, road construction
- Crowding
  - Subway, traffic congestion, pushing and shoving in any transit mode
- Weather
  - Heat, cold, humidity
- Environmental factors
  - Aesthetics, lighting, air pollution

Motor vehicles, along with other transportation modes, create "noise pollution" or simply noise which people find irritating and stressful. Obviously prolonged exposure to loud noises, including from motor vehicles and other transport modes, can be very stressful and impact adversely on hearing. Conversely,

> [I]n the context of transport, it is subtler physiological and psychological changes and their impact on human health and well-being that are generally of greatest concern. Some forms of noise, particularly regular exposure to sudden, loud noises and noise that disturbs sleep, may cause long-term and even permanent physiological changes, including the constriction of blood vessels and high blood pressure, both associated with cardio-vascular and circulatory disorders. Psychologically, noise may fuel aggressive annoyance and personal grievance, and also impair learning and workplace performance. (Hoyle and Knowles 1998:106)

Truck drivers may have accidents due to fatigue from long driving. Substance abuse is often a factor in many accidents incurred by truck drivers. Truck drivers may take drugs to relieve stress, but this coping mechanism may backfire and result in accidents and fatalities. Lawrence J. Quellett (1994:1–3) has written an interesting account of his experiences as a truck driver, noting that for him truck driving entailed periods of great stress, utter boredom, depression, and tremendous joy and satisfaction, or mixed-feelings in the proverbial sense.

Roadway Violence, Road Rage, and Drunk Driving

Peter J. Rothe (2008) presents a long litany of factors contributing to roadway violence:

- Social stress
- Alcohol consumption
- Social competition
- Unemployment and poverty
- Media culture
- Movies
- Video games which depict car blow-ups and crashes
- News stories
- Talk-back radio programs in which shock-jocks encourage aggression against cyclists
- Rap music
- Internet sites
- Corporate interests
- State-sponsored violence such as gas vans during the Nazi era and vehicles equipped with search lights, closed circuit TV, water cannons, public address systems, and psychological warfare systems

He also makes note of various other forms of violence involving motor vehicles, including wildlife poaching, road kills of animals, violence directed toward roadway workers (such as taxi drivers, bus drivers, and truck drivers), and drive-by shootings. Roadside memorials pointing to animal deaths incurred by collisions with motor vehicles exist in some places in the United States (Desmond 2013: 50).

Both experts and journalists have sought to explain road rage and aggressive driving. As Urry (2007:128) observes, "[w]ith 'road rage' emotions of aggression, competition and speed come to the fore although automobility is always polysemic, encouraging one to be careful and civilized *and* to enjoy speed, danger and excitement." Various studies have indicated that automobile drivers, particularly in conditions where traffic congestion is moderate to light, may exhibit "road rage" because they feel frustrated by their inability to reach their destinations quickly (Smith 2002). Road rage may be prompted by being cut off or tailgated by another driver, having to slow down by another driver, competing for a parking space, being honked at, and numerous other actions on the part of other drivers.

Road rage is culturally mediated in that it is expressed differently in different countries around the world: "[I]n Italy it is hesitancy that triggers noisy horn-blowing protest, in Germany it is slowness that triggers headlight flashing and tailgating, in Spain it is overtaking that prompts re-overtaking and sudden braking, in France it is being English and behaving like the French that necessitates vengeance, in Greece it is inappropriately following the rules that demands histrionic argumentation" (Michael 2001:69–70). A stationary form of rage may be directed toward parking service officers, as Hagman (2006:68) reports is the case in Swedish cities, who are "frequently subject to threats from angry car owners," to the point that the "sight of a bill on someone else's car may cause a car owner to burst out in rage."

Laurence H. Ross (1992:28–33) cites two main causes of "drunk driving" in the United States: (1) commitment to the use of alcohol and (2) commitment to the use of automobiles. While there is an element of truth to his claim, his analysis is reductionist in that it fails to identify political-economic and social structural forces driving these two predispositions. Ross (1992:119–127) proposes the following countermeasures based upon transportation policy to counteract drunk driving: (1) providing alternatives to driving: (2) planning for urban density that would reduce automobile dependence; (3) subsidizing alternative forms of transportation; (4) creating designated driver programs, which exist for example in Australia; (5) creating safe rides programs; and (6) reducing access to driving through various means, such as increasing the licensing age, establishing curfews on driving for younger drivers, driver training programs that stress the importance of restricting alcohol consumption before driving, and apprehending drunk drivers. In the United States, various organizations emerged to deal with road rage, aggressive driving, and drunk driving since the 1980s (Packer 2008:252). These include Mothers against Drunk Driving (MADD), Children against Road Rage (CARR), and Citizens against Speeding and Aggressive Driving (CASAD).

CONCLUSION

Chapter 5 examines the political ecological impacts of motor vehicles on physical health in terms of fatalities, injuries, and other health problems. It also discusses the political ecological impact of motor vehicles on mental health, including their role as an enabler of sociability and a source of stress, roadway violence, road rage, and drunk driving. Following Freund and Martin (1993:60), a political ecological approach to addressing the health consequences of automobility requires "changing the social and physical environment (e.g., building safer highways), producing safer cars, and making many alternative ways of traveling available to drivers." Automobile advocates in both the corporate and state sectors frequently assert that driving can be made safer by various means, such as requirements for seat belts and inflatable bags in the case of an accident, speed limits, road barriers, driver education, and drivers' licensing regulations. While particularly in developed societies, driving has indeed become safer, many motor vehicle-related accidents and fatalities still occur. However, in developing societies problems with motor vehicle accidents and fatalities not only persist but often are on the increase due to the growing level of motorization. Exposure to noise from road traffic causes stress, mood changes, sleep deprivation, fatigue, and anxiety. Motor vehicles emit pollutants, such as carbon monoxide, carbon dioxide, nitrogen oxides, hydrocarbons, particulate matter, and ozone that adversely impact respiratory health. Cyclists in particular because of their higher respiration rates are adversely affected when riding in traffic. Automobility has also contributed to a sedentary lifestyle, contributing to obesity.

Even the construction of safer highways and the production of safer automobiles will not be enough because the health benefits will be far greater by moving away from an infrastructure that perpetuates motor vehicle utilization. In this transition, cities must be redesigned to accommodate emergency and delivery vehicles and special vehicles to transport disabled people. Furthermore, the development of greener cities must be part and parcel of achieving social parity and will giving ordinary people more options in terms of their choices of where to live. The starting point for the de-automobilization of society is recognition of the automobile as a killer commodity: a cruel, mindless killer of health and environmental sanity. The automobile is not only a threat to physical health but also mental health and sociability, the latter two of which are intricately interwoven. Motor vehicle driving, whether it takes the form of driving an automobile, motorcycle, truck, or bus, while in some cases an exhilarating experience, is generally a stressful experience, particularly when roads are congested.

Automobility has resulted in widespread social isolation, either within the confines of the automobile itself or by promoting suburban sprawl and disconnectedness from shopping, recreational, community, educational, and health facilities. In seeking social connectedness, people in suburban areas are inclined to jump into a motor vehicle to access these amenities, rather than to simply walk or cycle to them or possibly take a train, tram, or bus to them. This applies to people of all age groups, but it is hardest on elderly people who no longer may have the physical requirements required for safe driving. Fortunately, in various developed societies, social services may provide vans to chauffeur elderly people from their residences to various amenities, but government cut-backs under the guise of austerity measures all too often have resulted in ending such programs. However, such amenities are non-existent or in their rudimentary stages in developing countries.

Chapter Six

Case Studies of Motor Vehicles in Various Places

This book provides illustrations of the historical development of automobility and the impacts of motor vehicles on settlement patterns, the natural environment, and health in many different places throughout the world. From an anthropological or even sociological perspective, what is needed is a cross-cultural and cross-national examination of motor vehicles, the beginning of which Daniel Miller (2001) promoted in his anthology on *Car Wars*. In developed societies, private automobiles, taxicabs, local and long-distance buses, motorcycles, motor scooters, motor bicycles, a wide assortment of trucks, and recreational vehicles account for the better bulk of motor vehicles. In developing countries, while most of these exist as well, one finds various other types of motor vehicles not generally found in developed societies. For example, the auto rickshaw is a generally motorized tricycle which is referred, depending on the national context, as a *three-wheeler*, a *samosa*, a *tempo*, a *tuktuk*, an *autorik*, a *mototaxi*, a *babytaxi*, or *lapa* (Verma and Ramanayya 2015:38). While, as Volti (1996:673) observes, an "obvious feature of American cars have been their size," automobiles and other motor vehicles in our countries, including European ones and Japan, tend to be smaller, often considerably smaller. In the spirit of contributing to a cross-cultural and cross-national examination of motor vehicles, in this chapter I present four case studies of automobiles in selected societies, namely West and East Germany, Australia, Brazil, and Cuba.

CASE STUDY #1—PEOPLE'S CARS IN WEST GERMANY AND EAST GERMANY: THE VOLKSWAGEN VERSUS THE TRABANT

Compared to the United States, Britain, and France, Germany up until the 1930s was slow to embrace the culture of automobility, largely due to the strong status of the coal-rail industrial complex. As Bonneuil and Fressoz (2016:116) report: "[I]n 1927 the ruling Social Democratic Party (SPD) chose to tax motor cars heavily in order to finance public transport. The creation of the public rail company Deutsche Bahn in 1920s, along with the municipalizing of the majority of tram companies, was likewise an element in a social policy aiming to reduce the cost of transport for workers." However, Hitler, who came to power in 1933, viewed modern nations as motorized ones and regarded the automobile as a "political resource and emanation of Germany's modernizing intentions" (Koshar 2005:122). Some 330,000 German citizens saved five marks a week to pay for a Volkswagen or what was termed "Strength through Joy." Hitler found a ready-made collaborator in his Volkswagen project in Ferdinand Porsche, a famous race car designer, who "had toyed with the idea of a small automobile since the early twenties" (Sachs 1992:59). He promised to manufacture a "vehicle that was economical to operate, with four seats for the whole family, air-cooled to avoid the problem of freezing winter temperatures, capable of a highway cruising speed of 100 kilometers an hour, and, above all, costing no more than 1,000 Reichsmarks" (Sachs 1992:60–61). Porsche signed a contract with the Nazi state on June 22, 1934, to build three prototypes of his concept of the people's car (Flink 1988:264). In late May 1937 the Nazi state-owned Deutsch Arbeitsfront (DAF) or German Labour Front provided the capital for the formation of the Gessellschaft zur Vorbereitung des Volkswagens (Volkswagen Development Company), which became in 1938 the Volkswagenwerk GmbH (Limited). While Porsche presented the first thirty test models of the Volkswagen in 1938, the 336,668 savers who had paid 280 million marks into the fund for their dream cars, never actually obtained them due to the war effort. The Wolfsburg factory, which Hitler hoped would be the "shining industrial center of his European empire," ultimately collapsed because of Nazi regime's need to transform the factory to a jeep production site beginning from 1940 on (Kerr 1993:85). Nazi Germany abandoned the Volkswagen factory and Wolfsburg to the approaching US Army on April 10, 1945 (Hoyle and Knowles 1998:321). However, the Americans turned occupation of Wolfsburg over to the British who took over operation of the Volkswagen plant, which with the assistance of many displaced persons, rolled out its ten-thousandth Volkswagen in October 1946. The British Military Government appointed Heinz Nordoff the general manager of the Wolfsburg plant and eventually on April 8, 1949, turned over *Volkwagenwerk* over to the German authorities. The Volkswagen factory essentially transformed the Kuebelwagen designed by Ferdinand Porsche into a civilian automobile (Ewing 2017: 10).

The West German masses only began to be able to purchase the broken promise of the Volkswagen in the aftermath of World War II, a period when Heinrich Nordhoff directed the company by focusing on the production of one model, which came to be widely known as the Beetle. Volkswagen complemented the Beetle with commercial vehicles, light trucks, the famous VW Microbus in 1950s, and the Karmann Ghia (a sports car on a VW chassis) in 1955 (Flink 1988:322). Returned servicemen there from the United States, UK, and British Commonwealth countries brought Beetles home with them, initially drawing animosity as "Hitler's car," but this turned into affection as Beetles proved their durability in world endurance trials in places such as outback Australia and Kenya. Parissien (2013:182) reports: "By 1960 VW production accounted for 42 percent of all cars made in West Germany, and VW became a public company. The firm had already been building homes in Wolfsburg [situated closed to the GDR border] for its plant's workforce since 1953, and in 1961 the Volkswagen Foundation was launched to conduct technological research. The same year, sales of the VW reached the five million mark, and a new, upgraded, 1500 variant was launched." About half of these five million Beetles, also called Bugs, had been exported around the world.

Volkswagen functioned as a state-owned firm until 1960 (Ewing 2017: 22). Under this arrangement, the federal government owned Volkswagen, but the state or *Land* of Lower Saxony controlled operation of the enterprise. When in 1960 the federal government opted to sell 40 percent of VW stock on the market, workers unsuccessfully attempted to thwart the sale but managed to obtain a concession. Workers controlled half of the seats on the VW board of directors but the Lower Saxony and federal governments also each had two seats on the board (Ewing 2017: 25).

By 1981 with 20 million Volkswagens sold, the Beetle stood as the most popular car in the world. The VW Beetle reportedly is the most produced automobile of all time (Patton 2002:2). Indeed, Volkswagen plants came to be situated "across the world, from Brazil to the Philippines to New Zealand (Parissien 2013:180). However, Kerr (1993:91) maintains that since the 1960s, "VW had never managed to repeat the international success of the Beetle, even with the Golf, which was the best-selling model in Europe."

Volkswagen experienced a serious dent to its international reputation when the news of its "clean diesel fraud" went viral around the world. Virtually overnight the world's largest automobile manufacturer and its association with quality, innovation, and reliability became replaced as a symbol of greed and deception. Due to numerous lawsuits and criminal investigations, Volkswagen by early 2017 had settled with regulators and automobile owners for $20 billion, with additional fines and claims still pending. Dan Carder, a researcher at the Center for Alternative Fuels, Engines, and Emissions (CAFEE) at the West Virginia University along with his research students first raised serious questions about VW emissions (Ewing 2017). In the United States alone, "[f]rom 44,000 in 2009, sales of Volkswagen diesels rose to 111,000 in 2013. That compared with fewer than 17,000 diesel passenger cars sold in 2007 by all manufacturers combined. Volkswagen was single-handedly responsible for a sixfold increase in diesel car sales in the United States" (Ewing 2017:150).

Although VW employees had detected serious flaws with their company's diesel emission system, VW management opted to conceal these, with company executives and engineers alike boasting about their supposed environmentally clean diesel technology. However, a small number of experts and activists in both Europe and the United States expressed skepticism about VW's boastful claims. While diesel motor vehicles emissions emitted less carbon dioxide than gasoline motor vehicles, they emitted far more toxic nitrogen oxides along with carcinogenic soot particles. Of course, Volkswagen by no means is the only European motor vehicle company which manufactures diesel vehicles. Diesel vehicles constitute about half of all vehicles sold in Europe and entail a lower fuel tax (Ewing 2017: 161).

The International Council on Clean Transportation (ICCT) awarded a very modest grant of $70,000 to Carder and his team to test various European diesel models and emissions technologies. The team obtained a 2013 BMW SUV and a 2012 VW Jetta from local rental agencies and borrowed a 2012 Passat from an owner. All three automobiles passed standardized tests at a California testing lab. On the open road, however, the "Jetta, the one with the lean NOx trap, belched unusually high levels of nitrogen oxides" while the Passat performed a little better and the BMW emissions control system performed as effectively as under lab conditions (Ewing 2017: 170).

VW sought to deceive California Air Resources Bureau (CARB) and Environmental Protection Agency (EPA) regulators lest the company hurt its market in the United States and probably elsewhere. Prior to the emissions device debacle, it had surpassed Toyota as the world's leading producer of automobiles. VW had captured a 21 percent share of the automobile market in China, surpassing General Motors, and was also strong in Latin America, particularly Brazil (Ewing 2017: 206). The company did not want to relinquish its hold on the US market, which included plans to manufacture a SUV at its Chattanooga (Tennessee) plant. On September 18, 2015, the EPA publicly accused VW of using software "that circumvents emissions testing for certain air pollutants" (quoted in Ewing 2017: 206) and revealed that it had discovered nitrogen oxide up to 40 times the limit in some VW vehicles, a defect that affected almost 500,000 automobiles in the United States. Ewing (2017:208) reports: "It soon became clear that Volkswagen was involved in one of the greatest corporate scandals ever. Between 2009 and 2015, the company had sold 11 million cars with software designed to deceive the emissions police. That included nearly 600,000 cars in the United States, 2.8 million in Germany, and 1.2 million in Britain."

In the aftermath of the revelations about the faulty diesel emissions device, Martin Winterkorn, resigned his position as the CEO of VW. However the VW management blamed nine employees for the problem and suspended them for their misdeeds (Ewing 2017:223). While VW incurred some hefty fines and a slump in its profits and ceased marketing diesel vehicles in the United States, the company went ahead with plans to manufacture the Atlas, a new gasoline-powered seven-seat SUV in Chattanooga, Tennessee (Ewing 2017: 273). Furthermore, Volkswagen has ambitious plans to add hybrid, electric, and autonomous automobiles to its repertoire of products.

Higher-status West Germans have generally opted to purchase other brands, such as the BMW or the Mercedes-Benz. Wilhelm Maybach designed the 1901 Mercedes for *Daimler Motoren Gesellschaft*, named for the elder daughter of Emil Jellinek, Daimler's leading agent in France (Flink 1988:33). Mass production of the Mercedes started up in 1904 at the new Daimler factory in Stuttgart, but fewer than 1,000 automobiles were being manufactured as late as 1909. The German Daimler company merged with the Benz firm to become the Mercedes-Benz company (Flink 1988:263). In December 1928 the Mercedes Benz was advertised at the International Automobile Exhibition in Berlin as the "car in a class by itself" (quoted in Sachs 1992:152). Much the same can be said of the Mercedes Benz today, although the BMW and the Porsche today would probably also qualify for the same characterization.

Perhaps spurred on by the Federal Republic of Germany's embrace of the culture of automobility, the German Democratic Republic did much of the same. Schmucki (2003:151) reports: "[T]he degree of motorization in both countries rose suddenly and rapidly after the war. It is a well-known fact that there were fewer cars in communist countries than in capitalist ones. What is more significant however is the fact that both East and West Germany experienced a similar growth rate in the number of cars. The East even saw a slightly higher rate of growth than the West."

As a major part of adopting the West German culture of automobility, the GDR's answer to the West German Volkswagen was the Trabant or *Trabbi*, a much more modest automobile to say the least. Gatejel (2016: 130–131) asserts: "The rising tensions of the Cold War and the harsh competition between the two German states increased the pressure on the GDR automotive industry to create its own car models." The GDR Politburo authorized the production of the inexpensive automobile, an East German counterpart to the popular West German Vokswagen in 1954 (Sandra 2010). VEB Sachsenring Automobilwerke Zwickau situated in the small city of Zwickau in Saxony began to manufacture the Trabant in 1957. Many elements of the automobile industry in Saxony that predated the war continued in the GDR, especially in Zwickau. These included widespread use of the two-stroke engine. The Trabant had a two-cylinder, air-cooled engine and a maximum speed of 100 kilometers per hour. Under the provisions of Comecon, the Soviet Union monopolized the production of larger automobiles, leaving the GDR with the Trabant and the somewhat larger and more luxurious Wartburg (Naegele 2004). In 1958 the Zwickau plant manufactured 1,750 Trabant P50s, in 1959 20,040 P50s and in 1960, 35,270 P50s (Rubin 2011: 128). From 1963 to 1991, the P601 functioned as the standard Trabant model.

East German customers had to wait up to 13 years after ordering a Trabant to take receivership of it. Due to a paucity of both automobile repair shops and spare parts, East Germans, particularly males, often became amateur automobile mechanics and engaged in *Autobasteln* or tinkering, which was not only a necessity but also became a recreational pastime (Moeser 2001). The GDR journal *Practic* and the *Motor-Jahr* [*Motor-Year*] carried articles on automotive tinkering and modification, allowing the self-made mechanics to make GDR cars, particularly Trabants, appropriate for camping and touring trips. Moeser (2011:164) observes: "Compared with those in the West, East German cars had lower annual mileages but nonetheless demanded a high number of hours devoted to maintenance and repair. Moreover, owners imposed pressure on themselves not to use the car they owned, or at least not on a daily basis, in order to save it from wear and tear. The time ratio between actual driving and working on one's car could become quite biased toward working." East German citizens sometimes obtained spare auto parts from their wealthier West German relatives. Automotive tinkering also occurred on vacation trips adjacent to one's tent and served as a ritual of solidarity as fellow campers observed and commented on this activity. Trabant owners installed special fog headlights and special wing mirrors and painted their cars in bright colors. During the GDR period, the Trabant was exported to Hungary, Czechoslovakia, Poland, Bulgaria, Romania, and Yugoslavia.

Trabants, along with Wartburgs, Soviet Ladas, Czechoslovakian Skodas traveled not only cities, towns, and on country roads but also on the GDR's autobahns, most of which had been built in the 1960s and the 1970s (Wolf 1996:101). In 1989 a version of the Volkswagen Polo engine replaced the two-stroke engine in the Trabant. Trabants transported many East Germans to Austria and West Germany in 1989, the former before the opening of the Berlin Wall and the latter after the opening. Beginning in 1990 and particularly after 1991, many East Germans abandoned their Trabants and purchased Western cars, often second-hand ones. Over the course of its history, VEB Sachsenring had manufactured 3,051,385 Trabants. People ordering Trabants generally had to wait 12.5 to 15 years before they received their car (Sandra 2010). Despite the Trabant's purported shortcomings, including having stiff seats and manual transmission that was difficult to shift, its two-stroke engine required less fuel than Western European automobiles. According to Freund and Martin (1993:72), the "average gasoline consumption in Eastern Europe in 1990s was 8.7 litres per 100 km, compared to Britain's use, which is close to the Western European norm, of slightly less than 10 litres per 100 km." In the late 1990s, an effort to revive Trabant production occurred in Uzbekistan, but only one model was manufactured. As part of GDR nostalgia, Trabant fan clubs have emerged throughout central Europe and the Trabant continues to be a symbol of Zwickwau. Rainer Eichhorn, Zwickau's mayor stated: "First of all, of course, the Trabant today, as in the past, really is part of our street life. And every year at the gathering of the international Trabant drivers [we hold] here—the last one attracted 30,000 visitors—one sees that the Trabant is not about to be killed off" (quoted in Naegele 2004). In the wake of the unification of the two Germanys, the Trabant production line was closed in 1991 and the Zwickau plant was purchased by Volkswagen AG. In an ironic twist of history, this was a case of the Western German "people's car" overtaking the East German "people's car." VW Sachsen managed to negotiate an agreement with the local trade union in which the latter agreed to lower wages and long working hours than those received by workers at the Wolfsburg plant (Greer 2008:187).

In addition, to the Volkswagen, many Germans, like many Americans, love their cars—all kinds of cars, particularly the ones that Germany manufactures, including the Mercedes-Benz and BMW, which are often driven at incredibly high speeds on the *Autobahns*. While Germany, in contrast to both the United States and Australia, is often depicted a *climate leader* rather than a *climate laggard*, due to its *Energiewende* or extensive program of decarbonization, particularly in terms of its commitment to wind and solar energy sources. Conversely, the "transport sector currently accounts for 163.6 Mt CO_2 corresponding to 20.5% of the country's emissions in 2015, and almost exactly the same amount as emitted by the sector in 1990 (164.4 Mt CO_2)" (Goessling and Metzler 2017:418). Furthermore, automobiles driven at speeds faster than 200 kilometers per hour increased from 7.6 percent in 1995 to 19 percent in 2006 (Goessling and Metzler 2017:420). Finally, newly German registered automobiles increased 3.6 percent in the period 2005–2014, in keeping with the Jevons Paradox.

CASE STUDY #2—AUTOMOBILES DOWN UNDER

Colonel Harry Tarrant produced Australia's first petrol automobile in 1897 (Lee 2010:229). In 1900 when Australia had a population of about 3.8 million people, rail was the primary mode of transportation in a huge country and trams serviced Sydney, Melbourne, Brisbane, Newcastle, Hobart, and Adelaide, but these scenarios were about to change. Most railway lines, including mining railways, had been built and operated by the government (Lee 2010:9). Trams paved the way for the suburbanization. They initially were imported primarily from the United States but later were manufactured in Australia.

Prior to the creation of Australia as a federal commonwealth in 1902, it functioned as a number of independent states and territories. Clarsen (2017:522) observes: "In Australia, resonances between national consolidation and automobility have been particularly strong. Automobiles arrived just when settler Australia was preoccupied with the creation of an imagined exemplary nation: white, prosperous and free from the class inequalities of Britain."

Australians are a highly mobile people for at least two reasons, the enormous size of a country with some 25 million people and the "tyranny of distance" they perceive with respect to much of the rest of the world, particularly Europe and North America. When I first came to Australia in 2004 on a one-year teaching stint at the Australian National University, I quickly became cognizant of many similarities between Australia and the United States, particularly California. In terms of automobiles, Paul Nieuwenhus (2014:82) asserts that "Australia has a strong and unique car culture, possibly rivalled only by that of the U.S." At any rate, Canberra, the national capital and a city at the time of some 325,000 people in 2004, had in large part been designed by Walter Burleigh Griffith, an American, in his design for modern Canberra laid out its streets in a series of concentric circles and roundabouts. Although Canberra has a reasonably good bus system, at least compared to US cities of a similar population, it is by and large a car-dependent city. Sydney, a city of some five million people, and Melbourne, a city of nearly five million people, probably more resemble the San Francisco Bay Area, which has some seven million, as opposed to Greater Los Angeles, a megapolis of some 15 million people. Like the San Francisco Bay Area which is served by a widely flung train system called BART (Bay Area Rapid Transit), both Sydney and Melbourne have relatively good suburban train systems, which have in recent years deteriorated, partly due to not keeping up with urban growth.

Automobiles were introduced into Australia in the early twentieth century and shaped urban planning very quickly afterwards, particularly in the case of Canberra, the new national capital. Robert Lee (2010:8) observes: "[C]ars were not part of everyday life for most city Australians until after 1950, although they were important in the bush and for urban rich by the 1920s. The adaption of cities to cars really occurred between 1950 and 1980."

When traveling between cities and towns during the late nineteenth century and the first half of the twentieth century, Australians tended to do so by train and even ferries, particularly in the case of Sydney and Melbourne. Indeed, ferries connected the Melbourne CBD with the suburb of Williamstown across Port Phillip Bay as well as Melbourne with various other towns around the bay, including Geelong, Queenscliff on the Bellarine Peninsula, and Sorrento on the Mornington Peninsula (Black 2010:178). Many Australian cities and towns also developed extensive tram systems which underwent evolution from being horse-drawn to steam and cable to electric-powered (Black 2010:183). Eventually in virtually all cities and towns, except for Melbourne, buses gradually began to displace trams beginning in the 1910s and 1920s. A rapid expansion in automobile ownership in Australia started during World War I with the introduction of the Model T Ford as well as Dodges, Buicks, and some other brands. In New South Wales alone, the number of motor vehicles increased from 3,978 in 1911 to 14,973 in 1916 (Black 2010:234). Between World I and World War II, motor vehicle companies with operations in Australia manufactured component parts. War bonds posters urged people to save for a post-World War II car (Davison 2004:3). However, World War II delayed the production of an automobile constructed from the ground up in Australia. As Davis (2016:184) observes, instead General Motors-Holden "built guns, armoured cars and boats, bombs, airframes and, eventually, Gipsy Major aero-engines."

General Motors Holden introduced a sedan as the first automobile completely manufactured in Australia in 1948 for which "thousands of Australians added their names to long waiting lists for the new car" (Loffler 2012:6). The company sponsored a tour of demonstration Holden sedans to country towns and rural districts in December 1948 (Loffler 2012:9). Automobile dealerships were encouraged to prioritize selling Holden sedans to taxicab companies. In 1951 GM Holden introduced the Holden utility, a small pickup truck (Loffler 2012:16). By the late 1950s, the Holden station wagon had evolved into a "mobile embodiment of a middle-class suburb family life" (Davison 2004:21). Like San Francisco, Melbourne has a tram system that serves the inner suburbs. Off course, it never developed a cable car system because, unlike San Francisco, its hills are not particularly high. The Bay Area, Sydney, and Melbourne all have widely flung bus systems, although of the three, I would say Melbourne's is the worst due to poor coordination with train and tram lines. Despite that the Bay Area, Sydney, and Melbourne have reasonably good mass transit systems, although many might debate this point, these cities on opposite ends of the Pacific Ocean have evolved into highly automobile-dependent cities, plagued by suburban sprawl and congestion during peak hours of travel. Flink (1988:130) astutely observes: "Australians, New Zealanders, and Canadians vastly preferred American cars over British makes, because American cars were cheaper, more reliable, and better suited to the primitive driving conditions that these countries shared with the United States. As early as 1912, the rugged Ford Model T had captured the Australian and New Zealand markets. By the mid-1920s, Ford faced stiff competition down under from GM's Holden operations, and in fact GM came to oversell Ford in Australia."

Ford Australia manufactured automobiles in its Victorian plants in 1950s according to the same design as Fords produced in the UK (Lee 2010:315). Chrysler initially manufactured its automobiles using an imported chassis but became the third automobile company to develop full design and manufacturing operations in Australia (Lee 2010:317). It opened a factory at Tonsley Park near Adelaide where it manufactured the Valiant. In the 1970s, the Holden, Falcon, and Valiant became known as the Big Three in Australia. Australian Motor Industries based at Port Melbourne assembled the first Toyota automobile built outside of Japan in 1963 (Lee 2010:320). Mitsubishi and Nissan also produced automobiles in Australia but shut down their operations around 2008. Currently, Ford has an assembly plant in Broad Meadows (Melbourne) and had an engine plant in Geelong, which closed in 2016, both in the state of Victoria (Australian Government 2008:11). GM Holden manufactures motor vehicles in Elizabeth (Adelaide area) and V8 engines at Fisherman's Bend (South Melbourne). Toyota manufactures motor vehicles and four-cylinder engines in Altona (Melbourne). Export of Australian motor vehicles increased from 16 percent of local production in 1997 to 42 percent in 2007, with the Middle East serving as the primary export market (Australian Government 2008:15).

Since the early twentieth century, like the other Anglophone countries of the United States, Canada, and New Zealand, Australia has become an automobile society par excellence. In March 2010 Australia had 16.1 million registered motor vehicles, comprising 12.2 million passenger vehicles, 2.5 million light commercial vehicles, 430,000 rigid trucks, 82,000 articulated trucks, 86,000 buses, and 660,000 motorcycles (Australian Bureau of Statistics 2010). Whereas in 1955 there were 153 passenger vehicles per 1,000 people in Australia, this rate had increased to 568 per 1,000 people by 2013 (Australian Bureau of Statistics 2013).

Motor vehicles played a central role in connecting a relatively small population of barely four million people, particularly in rural and outback areas.

To some degree, even Indigenous Australians, despite their socioeconomic marginality and the fact that they were disenfranchised from voting until 1967, participated in this process. As Clarsen (2017: 527) observes, between World I and World II, "Aboriginal people began to use automobiles to maintain cultural practices; to earn money, for mobile housing; to stay in touch with scattered kin; to organize politically; to keep one step ahead of welfare authorities; to provide dignified transport for funerals, confinements, and weddings; for picnics, holidays and shopping trips; as a buffer against racism on the road; and for medical emergencies."

Like in other countries, such as the United States and the UK, in the early twentieth century, competition for access to the city's streets in Australia exhibited a touch of class struggle. Davison (2016:161) reports: "Cars were 'rich men's toys,' beyond the reach of everyone except the wealthiest five or ten per cent of the population. Their drivers, mostly men, regarded them as an extension of male power and authority. They resented sharing the road with slower-moving horses and buggies, obeying outdated traffic laws and submitting to the authority of police officers, who were mostly recruited from the working class." By 1914 there were an estimated 37,000 motor vehicles consisting of over 100 models operating in Australia (BMC-Leyland Australia Heritage Group 2012:22). In terms of motor vehicle densities, Australia with 127 people per motor vehicle trailed the United States with 77 people per motor vehicle, but it was ahead of the UK with 165 people per motor vehicle, France with 318, and Germany with 950.

The rise of the automobile in Australia posed a serious threat to publicly owned railways. While initially both conservative and progressive politicians promised to protect the railways, lobbying on the part of the motor vehicle-highway complex quickly altered the tide. The National Roads Association established in Sydney in 1920 lobbied for the formation of a board to manage roads in New South Wales (Lee 2010:238). The National Roads and Motoring Association and the Automobile Club of Victoria (later the Royal Automobile Club of Victoria or RACV) also fiercely lobbied for road construction (Davison 2016:163). The Victorian government created a Country Roads Board with the aim of ensuring farmers that they would have access to deliver their products to the state's railway stations and constructed the Great Ocean Road from Torquay to Warrnambool, a project that greatly stimulated tourism.

The Ford Manufacturing Company of Australia opened a plant for manufacturing and building motor bodies and parts in Geelong, Victoria, in 1925 (Conlon and Perkins 2001:29). The Ford Motor Company of Australia assumed responsibility for assembling and distributing motor vehicles. T.J Richards & Sons began to build bodies for chassis imported by the South

Australian distributors of the Chrysler Corporation's Dodges in 1926. General Motors set up its manufacturing facility under a contract with the Holden Motor Body Builders at Fishermen's Bend near Port Melbourne in 1926 as well. It took over Holden in 1931 when the latter company faced insolvency due to the Depression, resulting in General Motors-Holden's Limited (GM-H) (Conlon and Perkins 2001:29). US automobile companies dominated the Australian market during the 1920s. In 1927, 81 percent of new automobile purchases were of US models, as opposed to only 14 percent of UK models and five percent of Continental European models (BMC-Leyland Australia Heritage Group 2012:23).

Automobile density in Australia increased from one car for every 55 people in 1920 to one car for eleven people in 1929 (Lee 2010:236). The Great Depression starting in 1929 dampened the expansion of the Australian automobile industry. Because US automobiles had increased in both size and cost during the 1930s, British cars temporarily penetrated this market. Nevertheless, many urbanites still relied on public transit. When Perth abandoned its trams in 1933 and Sydney in 1934, trolley buses picked up the slack for a while, but eventually were abandoned (Lee 2010:199–200).

Like other Allied powers, the Australian government-imposed petrol rationing during World War II starting in 1940, a policy that was later intensified, meaning that by 1942 the individual allowance was roughly equivalent to 800 miles of driving per vehicle per annum (Glover 2011:3). The government relaxed petrol rationing after 1944 and terminated it in 1950, but at this time automobile ownership in Australia remained by and large an "elite affair until the 1950s" (Lee 2010:226). In 1951, the Melbourne and Melbourne Board of Works (MMBW) commissioned a travel survey which "found that car ownership rates in Melbourne were still low: the city average was 121 cars per 1,000 residents, with rates ranging from 61.5 in inner Melbourne to 183.7 in the wealthy Malvern-Caulfield subregion" (Stone and Mees 2016:114). The Opinion Research Centre (1951:115) reported that whereas only 19.5 percent of people traveled to work in a private motor vehicle, 26.0 percent did so by train, 22.1 percent by tram, 8.8 percent by bus, 9.5 percent by bicycle, and 14.1 percent by walking. The advent of the age of mass automobility in the 1950s quickly altered such commuting patterns in Melbourne and other Australian cities.

The new Holden Australian automobile went into production in early 1949 and became very popular because it was advertised as "The Australian car" and filled a space between larger and more expensive American automobiles and the less expensive, smaller, and less powerful British automobiles (BMC-Leyland Australia Heritage Group 2012:27). The Holden accounted for about half of the Australian automobiles for cars in its range of models by the mid-1950s (Fleming, Merrett, and Ville 2004:92). ALP Prime Minister Ben Chifley presided over the launch of the Holden at the GM-H headquarters at Fishermen's Bend in Melbourne. John Wright (1998: ix) asserts that the history of the Holden is a "metaphor for the history of postwar Australia." GM-H ironically also manufactured trams for the South Australian government and the Melbourne Metropolitan Tramways Board for a while (Wright 1998:1). The Holden was first exported to Thailand, Malaya, and North Borneo in 1956 and the first Holden station was produced in March 1957 (Wright 1998:79).

Japanese automobile companies had operations in Australia for some time. Nissan purchased the Motor Producers facility in Dandenong, an eastern outer suburb of Melbourne, where Volkswagens were also manufactured for over a decade (Minchin 2007:331). It reportedly produced about 50,000 vehicles per annum in the late 1980s and accounted for 12.9 percent of the Australian automobile market in 1990.

The Australian state at both the federal and state levels, particularly after World War II, played a key role in promoting the culture of automobility. Dodson and Sipe report:

> The Commonwealth encouraged a domestic motor industry—it would be a symbol of Australia's industrial vitality. Public transport networks had become run down, as a result of high demand and low investment during the Depression and World War II. The new outer suburbs had not been provided with extended tram services or new rail infrastructure. Instead, grandiose freeway schemes, from an even more suburbanising United States, were inveighing their way into state metropolitan plans. (Dodson and Sipe 2008:16–17)

The Australian government also gave strong support to the production of the first Holden (Moriarty and Beed 1992:30). During the 1950s and 1960s government-sponsored tariffs attracted various foreign motor vehicle companies to establish manufacturing facilities in Australia. By 1983 when the Hawke ALP government assumed power, "the nominal tariff on imported cars and components had reached 57.5 percent" (Lansbury, Saulwick, and Wright 2008:13). In subsequent years, tariffs on imported motor vehicles declined significantly under both ALP and Coalition governments.

The Australian government and state governments have at times provided generous subsidies to the motor vehicle industry. For example, the Australian government, with some assistance from the Victoria government, provided Ford Australia with a grant of $52.5 million so that it would develop the new generation Falcon and design and engineer a pick-up truck platform intended primarily for export (Australian Government 2008:34). The Australian government, along with the Victorian and Southern Australian governments, gave a $6.7 million grant to GM Holden so that it would "introduce safety and fuel management improvements and further reduce greenhouse gas emissions from Commodore vehicles" (Australian Government 2008:34).

In 1900 when Australia had a population of about four million people, Sydney, Melbourne, Brisbane, Adelaide, Hobart, and Newcastle had well-used tram systems. Of these cities, only Melbourne today has a relatively extensive tram system, although Adelaide has a light rail line that connects its CBD with coast suburb of Gleng (Laird and Newman 2001). Australian cities developed as regions of urban and suburban sprawl, committed to the notion that each household should own a quarter-acre block and one or more cars to transport its occupants through the resulting sprawl. Car-dependency is especially acute in outer suburban areas.

At the end of World War II, Melbourne's electric trains transported some 190 million passengers per annum, with government-owned trams and buses carrying an additional 355 passengers (Beed and Moriarty 1988:57). Automobile infrastructure is politically constructed: "Powerful interests have made their influence felt in lasting ways, notably the construction of roads" (Ladd 2008:8). In Australia, the motor vehicle lobby and the Liberal Party eventually convinced the Australian Labor Party (ALP) to join them in the promotion of road construction rather than prop up a deteriorating public transport system (Davison 2004:125). Planning for Sydney's freeways started in 1943 (Lee 2010:111).

Despite a doubling of Melbourne's population, in 1985 train ridership had decreased to 86 million and trams/buses to 140 million. Davison (2004:261) notes that "by the 1990s more than 80 percent of daily journeys in Melbourne were made by car, and more than 50 percent of households had two or more cars." Indeed, every city in Australia, except Melbourne, lost its trams and trolley buses between 1950 and 1970 (Pointing 1992:337). Perth and Adelaide removed their tram lines in 1958 as did Sydney and Brisbane in the 1960s.

The motor vehicle lobby easily persuaded the Liberal Party and eventually convinced the Australian Labor Party to promote road construction rather than improve public transportation. In his role as Victorian Premier, Henry Bolte strongly supported motorway construction (Curtis and Low 2012:23). In his vision of Brisbane as an ideal city, Australian Labor Party Lord Mayor Clem Jones envisioned a scenario in which every working person could own a car, which contributed to ultimate demise of the city's tram system, which a group of North American consultants recommended in 1965 (Lee 2010:203). In the 1960s, the Australian Commonwealth or federal government developed a national roads policy that eventually resulted in the creation of a national highway system. Local governments have responsibility for local roads and the states for main roads. Gray (2011:31) reports: "The states provide 48 percent of total government and road funding, local government 29 percent and the Commonwealth 22 percent. However, the main sources of funding for interstate highways and a significant contribution to local roads are controlled from Canberra. The Commonwealth still works with the states, which provide the largest share of funding and administer their own work and oversight of local government."

Australia relied heavily upon American highway planning, in part fostered by the fact that Australian engineering students were provided with scholarships to enroll in the traffic engineering course at Yale University. Furthermore, during the 1960s, "Australian governments hired American road engineers to draw up extensive freeway blueprints inspired by Los Angeles and Detroit" (Dodson and Sipe 2008:17). Davison (2016:172) reports: "During the 1960s, each of the Australian capital cities carried out major transportation studies, based on American models usually with the advice of American consulting firms. The 1961 review of Sydney's freeway plan, the Brisbane Transportation Study (1965), the Melbourne Transportation Study (1966–69), the Metropolitan Adelaide Transportation Study (1968), and the revision of the De Laue Cather Perth Study (1971) all projected massive increases in traffic volume and laid out ambitious freeway plans."

Despite state support for road construction, road service agencies in Australian cities consistently argue that state governments do not provide sufficient funding for road construction (Curtis and Low 2012:9). Nevertheless, Brisbane implemented a freeway plan in 1965 on the north shore of its river and Melbourne initiated a freeway plan in 1969 that eventually resulted in the South Eastern and Tullamarine freeways, City Link and the West Ring Road. The ALP government under the leadership of Prime Minister Bob Hawke promoted highway development over rail development (Laird and Newman 2001:9). According to Dodson and Sipe (2008:19), "Governments in most states found new friends in private finance to fund freeways, further embedding the car as the essential mode of suburban transport. And the Commonwealth maintained a significant urban road funding role through its Auslink program. Many of the new roads have allowed private 'roadlord" companies to extract tolls from suburban motorists: Sydney residents, for example, have seen toll road systems in their city."

Automobile dependence of the least wealthy people rose between 2001 and 2006, indicating that there is a strong relationship between automobile dependency, low incomes, and mortgage stress in the outer suburbs of Australian cities (Dodson and Sipe 2008).

For some three decades in the wake of World War II, Australian governments imposed high tariffs, up to 57.5 percent in the early 1970s, on automobile imports, thereby protecting motor vehicle manufacturing operations based in Australia (Auer, Clibborn, and Lansbury 2012:147). Dennis (2016:26) reports: "In the 1980s, federal Labor senator John Button's Motor Industry Development Plan transformed the industry into an outward-looking sector that was, for the first time, export oriented and innovation focused. Where industry policy had for decades been designed to protect Australian-made cars from imports, Button's plan rewarded manufacturers who could export their cars or components." The Australian government gradually began to reduce tariffs on imported cars, to a low of five percent in 2012, thus disadvantaging domestic motor vehicle production. This, for instance, contributed to the closure of the Nissan plan in Melbourne in 1992. By 2007 Australia had one of the lowest motor vehicle tariff rates in the world, ten percent for private motor vehicles and five percent for good transport vehicles (Australian Government 2008:38).

In March 2008 Mitsubishi Motors ceased its Australian production operations (Australian Government 2008:11). To preserve what remains of a domestic motor vehicle industry, Australian governments at both the federal and state levels have provided generous direct subsidies to the Australian automobile industry. For example, in 2009 the Australian government provided Ford with $38 million for its proposal to develop a fuel-efficient EcoBoost engine for use in its Falcon model (The Allen Consulting Group 2013:7). In a similar vein, when Holden announced in March 2012 that it would manufacture the Holden Cruze and another model at its Elizabeth facility in South Australia, the Australian, South Australian, and Victorian governments provided a $275,000 assistance package. By the beginning of the twenty-first century, the Australian motor vehicle industry consisted of several hundred component supplies and four foreign-owned assemblers, namely Ford, GM-Holden, Toyota, and Mitsubishi (Lansbury, Saulwick, and Wright 2008:13). Minchin (2007:341) reports: "All five Australian manufacturers were subsidiaries of foreign firms and faced a constant battle to persuade their parent companies that they deserved investment. In the late 1980s and early 1990s, the profitability of Australian car producers was patchy; from 1985 to 1993 there were only two years of overall profits for the manufacturing options of the big five."

GM Holden has only one plant now, situated in Elizabeth, South Australia. Whereas Holden sold some 95,000 Commodores in Australia in 1998, it sold fewer than 30,000 in 2015 (Davis 2016:186). The Australian automotive industry in the early twenty-first century was a struggling one, despite it being situated in a highly automobile-dependent country, one which finally more or less totally collapsed in 2016. In 2007 the Australian automobile industry accounted for 5.6 percent of manufacturing value-added but only 0.6 percent of Australian GDP (Conley 2009: 222). In contrast to 1990 when locally manufactured automobiles accounted for 68.9 percent of Australian sales and 41 percent in 2000, this figure had fallen to 24.7 percent in 2007, largely due to a drastic decline in automobile tariffs from 57.5 percent in 1984 to ten percent in 2005 and then five percent in 2010 (Conley 2009: 222).

A 2014 Australian Parliamentarian research paper reported that in 2013–2014 the contribution of the Australia's motor vehicle manufacturing industry to the "gross domestic product (GDP) was 6.5 per cent, which is less than half what it was four decades earlier" and showed no sign of reversing this trend (quoted in Davis 2016:180). Given the tyranny of distance, imported automobiles coming from Japan but particularly the United States and Europe must be shipped long distances. The Australian automotive manufacturing consists of subsidiaries of Ford, General Motors Holden, and Toyota "as well as hundreds of parts manufacturers, ranging from small Australian producers to companies that are also subsidiaries of very large multinationals, such as Bosch" (The Allen Consulting Group 2013: vii). At least in 2013, it employed some 50,000 people, of which about 17,000 worked at Ford, General Motors Holden, and Toyota. At the time, the Australian automotive industry manufactured about 225,000 motor vehicles, down from a historic high of 407,000 in 2004. Mitsubishi closed its Adelaide plant, resulting in sacking thousands of workers (Dodson and Sipe 2008:43). When the Coalition government under Prime Minister Tony Abbott refused to "commit to continued budgetary assistance, which in 2013–2014 fell to its lowest level in several decades, Ford, GM, and Toyota announced plans to close their Australian production operations (Clibborn, Lansbury, and Wright 2016:7). Ford closed its Geelong (Victoria) plant in 2016. At its peak Holden had a workforce of 30,000 spread out over ten production facilities but now employs about one sixth this number at two facilities (Haigh 2013:7). Holden along with Toyota closed all their Australian automobile factories in 2017 (Denniss 2016:24). Free trade agreements have had a negative impact on the Australian automobile industry. A case in point is a free trade agreement between Australia and Thailand: Standard (2016:4) reports: "One sector experiencing especially rapid import penetration from Thailand has been the vehicle sector (Australia imported over $6 billion in automotive products from Thailand last year and exported almost nothing—just $37 million—the other way). The surge in automotive imports from Thailand undeniably contributed to the decisions by global automakers to stop operations in Australia."

Australia imposed motor vehicle emissions standards for new vehicles in the early 1970s and these have improved over time, with leaded petrol being phased out in late 2001 (Kelly and Donegan 2015:16). Conversely, fuel efficiency in the Australian automobile fleet, most of which now consists of imported models, has not been impressive and has not improved significantly over time. Whereas fuel efficiency increased from 11.4 liters/100 kilometers in 1963 to 12.6 liters/100 kilometers in 1976, by 2006 it had returned to 11.4 liters/100 kilometers (Mees 2010:43). Motor vehicle manufacturers in Australia "have long lobbied to keep Australian emissions legislation below best international standards, saving them large amounts of money" (Davis 2016:196).

As is the case in North America, automobile and provision of roads and parking spaces have dominated transportation planning in Australia since 1960s (Stone and Beza 2014:91). Of the capital cities, this was particularly true of Melbourne, which experienced an increase of 14.9 percent in automobile driving as opposed to a decline of 8.3 percent in public transportation and a decline of 2.6 percent in walking between 1976 and 2011 (Groenhart and Mees 2014:125). This decline can be attributed to the construction of freeways and tollways, the failure to have not constructed a new suburban rail line since 1930, a poorly coordinated bus transit system, the deterioration of tram service due to congestion and reductions in frequency of runs. The revival of the Melbourne CBD beginning in the 1990s resulted in over 28,000 vehicles a day on Swanston Street, a principal arterial connecting the CBD with the University of Melbourne and Carlton to the immediate north. Fortunately, most of this arterial became closed off to most motor vehicle traffic several years ago but continues to be used by trams and bicycles (Adams 2009: 43). Whereas the short-lived Whitlam government of 1972–1975 sought to transfer railways from the state governments to the federal government, the conservative state governments in Victoria, New South Wales, and Queensland resisted this proposition (Troy 2010:177). Although American Walter Burley Griffith's original design for Canberra included suburban train and tram lines, by the 1950s these plans went by the wayside, thus ensuring that the federal capital would be by-and-large a car dependent city (Mees and Groenhart 2014:184).

Like elsewhere, roadways in Australia have been a bone of contention among various stakeholder groups. The pro-road faction is represented by groups such as the Royal Automobile Club of Victoria, the Royal Automobile Club of Western Australia, and the National Roads and Motorists' Association in Sydney. Road service agencies in Sydney, Melbourne, Perth, and presumably other Australian cities repeatedly complain about what they view as inadequate funding for roads (Curtis and Low 2012:9). The pro-public transport faction being represented by groups such as the Public Transport Users Association, Eco-Transit in Sydney, and the Transport Action Coalition in Perth (Curtis and Low 2012: 155, 167). The Rapid and Affordable Alliance, which consists of 17 environmental, health, and union groups (including the Australian Council of Trade Unions and the Australian Conservation, published a report on June 23, 2009, entitled "Investing in sustainable energy: our clean, green transport future." This report recommends that the federal government devote two-thirds of transportation funding to public transit and only one-third to roads during the next few decades. It noted that between 2004 and 2009 the federal government had spent only $1.2 billion on trains but $14 billion on roads. Despite this, ridership on public transportation is now increasing faster than automobile use per capita in some Australian cities, with the highest growth being on suburban rail in Perth, suburban trains and trams in Melbourne, and suburban trains and buses in Brisbane (Newman and Kenworthy 2015:18). A light rail system was completed on the Gold Coast of southeastern Queensland in 2014. Construction on a light rail system in Canberra has been planned, long overdue given that the fact that Canberra has been for long well-positioned for such a system. The Link-Up Conference and other similar events have emerged as forums for anti-freeway groups in Australia.

Despite some moves to shift away from automobile dependence to greater reliance on public transit, the automobile continues to constitute a central feature of Australian sociocultural life. Australian Bureau of Statistics (2016) reports that for purpose of commuting to work, 68.7 percent drive an automobile, 5.1 percent ride in an automobile as a passenger, 0.7 percent ride a motorcycle or motor scooter, 0.7 percent drive a truck, 0.2 percent ride in a taxi, 5.1 percent travel by train, 3.4 percent travel by bus, 0.1 percent take a ferry, 1.1 percent cycle, 0.8 percent travel by some other means, 3.9 percent walk, and 4.2 percent commute by multiple means. The 2011 Census of Population and Housing revealed that 87 percent of Tasmanians, 85 percent of Queenslanders, and 83 percent of Australian Capital Territory (mainly the city of Canberra) residents use a passenger vehicle to commute to work (Australian Bureau of Statistics 2013). Sydney is unusual in that it has an extensive ferry system that plies Sydney Harbour and the Paramatta River, although Brisbane has a more modest ferry system that plies the Brisbane River. While Canberra is roughly on par with most other Australian capital cities in terms of car dependency, an extensive bicycle lane network makes it the capital city with the greatest percentage of people who commute to work by cycling.

Reliance on automobiles in Australia is an expensive undertaking. In a recent survey, the Australian Automobile Association (2016:5) reports that in a hypothetical household consisting of a couple with two automobiles, the "typical Australian passenger vehicle is driven 13,000 kilometres per year." In the second quarter of 2016, the typical household spent 16.8 percent of its weekly income on transportation in Sydney, 14.1 percent in Melbourne, 15.9 percent in Brisbane, 10.1 percent in Perth, 13.2 percent in Adelaide, 14.2

percent in Hobart, 12.0 percent in Darwin, 10.5 percent in Canberra, and 13.3 percent nation-wide (Australian Automobile Association 2016:7). Most of these expenditures went into ownership, operation, and maintenance of a private motor vehicle rather than utilization of public transit.

As an interesting aside, New Zealand, another "down under country" situated across the Tasman Sea from Australia, is even more automobile dependent than its larger neighbor. It has the highest number of motor vehicles per capita of all OECD countries. Hopkins (2016:152) reports: "New Zealand's geography (e.g. long, thin country, spatially dispersed urban centres), low population density, and high levels of urbanization also contribute to distinct cultures of mobility which prioritise private motorised modes. Both inter and intra city travel is reliant on private transport modes, with low investment in public modes such as rail and bus." New Zealand does have a train line that connects Auckland to the national capital of Wellington on its North Island and train lines that connect the north shore of the South Island with Christchurch and the latter with Greymouth on the Tasman Sea. New Zealanders commonly fly between cities and towns on the two islands.

CASE STUDY #3—BRAZIL

One of the principal reasons that Latin America has become dependent upon motor vehicles of various sorts, particularly automobiles, buses, and trucks is that "with a few luxurious exceptions, trains are often old and less than reliable" (Sernau 2009: 315). Brazil conforms to this pattern. Brazil entered the automobile age by importing 24,475 cars between 1902 and June 1920 (Wolfe 2010:26). In 1920 Brazil had one automobile per 1,400 people, superseded by considerably more affluent Argentina which had one automobile per 281 people but slightly ahead of Venezuela which had one automobile per 1,421 people but far ahead of Bolivia which had one automobile per 5,454 people. The first motor vehicles to penetrate Brazil's interior were trucks which transported sugar cane and coffee from the plantations to railway lines during the 1920s and 1930s (Wolfe 2010:45). Furthermore, few roads connected Brazilian cities with one another. After World War II, the Brazilian government, however, built highways between Sao Paulo and Rio de Janeiro, Sao Paolo and Belo Horizonte, and Sao Paulo and Curitiba.

Although many Latin Americans expressed antipathy to the penetration of their countries by large US multinational corporations, Wolfe (2010:67) maintains that Brazilians from a broad spectrum of society welcomed the installation of Ford and General Motors factories in their country. At the same time, US motor vehicle companies lobbied for the expansion of Brazil's highway system which in turn spurred the demand for even more automobiles, trucks, and buses (Wolfe 2010:78). They also promoted the idea that women drivers constitute an essential component of modernity. Ford dealers promoted their tractors as part and parcel of industrial agriculture that would contribute to Brazil's economic development. President Vargas promoted the creation of Brazil's first national traffic code in order to facilitate the country's culture of automobility (Wolfe 2010:102). Beginning in the 1920s the creation of bus transportation in Sao Paulo permitted working-class people and poor people to build houses as well as shantytowns on the periphery of the city (Wolfe 2010:158). Furthermore, the number of automobiles in Sao Paulo climbed from "about 63,000 in 1950s to 415,000 in 1966, to 1.4 million in 1993" (Wolfe 2010:159). President Juscelino Kubitshek during his five-year term between 1956 and 1961 "oversaw the creation of a domestic automobile industry through the broad entry of foreign investment, the building of highways that connected distant regions, and the construction of Brasilia, the car dependent, modernist national capital in the interior" (Wolfe 2010:113). He viewed automobiles as an important instrument in nation-building and predicted they would eventually facilitate movement from coastal areas to the interior. Ford and General Motors also expanded their manufacturing operations to the production of trucks and buses at their Sao Paulo factories.

Janio Quadros and Joao Goulart served short terms as controversial presidents between early 1961 and April 1964, prompting business people, military leaders, and the Lyndon B. Johnson administration to mount a coup which resulted in a military dictatorship ruling Brazil for 21 years. In the mid-1960s the government built a highway connecting Brasilia and Belem, thus providing a link between the southern and northern regions of the country. Wolfe (2010:153–154) reports: "Military planners imagined a series of modern, paved highways that would finally fulfil the Amazon's promise as a major source of national wealth and pride. Economic development would follow the Transamazonian Highway, providing opportunities to migrants who would settle on small plots and larger agribusinesses that would be able to more efficiently exploit the land. . . . Beyond the stated need to develop the interior, the military regime was also interested in creating a stable population and providing a safety valve for urban discontent." In addition to Ford and General Motors, numerous foreign motor vehicle companies prospered during the era of the military dictatorship, transforming both Brazilian cities and much of the areas in between into an elaborate road system highly dependent on automobiles, buses, and trucks. Nieuwenhuis reports that in Brazil,

> By the 1960s the local car industry was dominated by Volkswagen, DKW-Vemag, a German Brazilian combine boasting some local designs, Simca (France/US), Willys-Overland (American Motors) and Alfa Romeo. Later

Toyota, GM and Chrysler played a role, Ford made a Renault-derived car, while Fiat, a late entrant in 1988, rapidly became a dominant player in the 1990s. Brazil developed some local designs, some of which (Puma, Volkswagen SP2 and VW Brasilia) enjoyed some exports. . . . The truck industry is dominated by Swedish firms Volvo and Scania, with VW a secondary player. (Nieuwenhuis 2014:46–47)

During the reign of the generals, the vast majority of Brazilians could not afford to purchase an automobile. Automotive workers, even during the reign of the generals, periodically have engaged in strikes to obtain better wages and working conditions. As a result, they became a labor aristocracy in that they "were among the best compensated workers in Brazil" who fully embraced the culture of consumption, including ownership of automobiles. Under the new democratic system that emerged in Brazil in the late 1980s, a former autoworker and trade unionist Luiz Inacio Lula da Silva became a founding member of the Workers' Party (PT), and was elected to Congress in 1986. Finally, following three unsuccessful bids, Lula was elected to the presidency in late 2002 and re-elected in 2006 (Wolfe 2010:181–182). The iconic status of automobiles in modern Brazil is testified by the fact that Brazilian Ayrton Senna, a three-time winner of Formula One world championship who died a few hours after his racing car crashed at the Grand Prix in Italy on May 1, 1994, received a state funeral in Sao Paulo (Wolfe 2010:3). Over the course of the twenty-first century, Brazil has emerged as the only Latin American country in the BRICS consortium, the other countries being Russia, India, China, and South Africa, that some scholars refer to as "sub-imperialist" states (Bond 2015). Luce (2015:37) reports that "[u]nder the aegis of the diversified industrial model, Brazil has become the leading Latin American automobile producer and the ninth in world ranking."

Sao Paulo, the largest city in Brazil with a population of over 20 million people and one of the largest cities in the world, has become highly congested due to a seemingly never-ending increase in motor vehicles, particularly private ones. Eriksen reports:

On a Friday afternoon, the sum of the length of the traffic jams in and out of the city often reach more than 200 km. On a Friday evening in November 2011, a total of 500 km of traffic jams was measured in the city, minor roads included. It may well be a world record. Paulistas are proud of the metro, but the lines extend only for 75 km. Many car drivers spend the time in the car watching TV, applying make-up, shaving, talking on the phone or sending email, while watching the bumper ahead. A lawyer who spends between two to three hours getting to work every day, says that he feels like a prisoner. (Eriksen 2016: 82)

Although most large Brazilian cities have adopted automobility full-on, Curitiba has provided a viable alternative to it in the form of a rapid bus transit system when the city's leaders in the 1960s adopted urban planning inspired by Donat-Alfred Agache of the French Society for Urban Studies before its streets could become flooded with automobiles (Wolfe 2010:161). Curitiba adopted a master plan based upon the ideas of Jamie Lerner, an architect based at the Federal University in the city. "It closed some roads to motor vehicles and created a series of pairs of one-way streets to move traffic in opposite directions. The city also created a complex system of buses that reduced congestion and rationalized movement throughout Curitiba. The city continued to update its transportation system in the 1980s with the implementation of the Integrated Transportation Network . . . which relies on a bus rapid-system with so-called Speedybuses"(Wolfe 2010:161). The Speedybuses are articulated vehicles which can accommodate 270 passengers and utilize large covered bus stops. Only time will tell whether other Brazilian cities will embrace a more sustainable transportation system as Curitiba has.

In contrast to Curitiba is Sao Paulo, a mega-city which underwent an increase in freeway length per person of 111 percent between 1996 and 2011 (Newman and Kenworthy 2015:89). In comparison to North American, Australian, and European levels, Sao Paulo has modest levels of automobile and motorcycle ownership (365 and 73 per 1,000 persons respectively in 2011, but high in comparison with most cities in developing countries. Newman and Kenworthy report:

> The amount of public transit service is very high by global standards and is increasing significantly (vehicle- and seat-kilometres per capita rose by 32 percent and 21 percent respectively, over 15 years). While buses remain clearly dominant within the transit system (in 2011 they constituted 75 percent of all public transit vehicle—kilometres in the region), their growth in numbers is faltering compared with that of rail, and the region is moving toward a much greater role for its metro and suburban rail systems (bus use in terms of passenger-kilometers per capita declined by 4 percent, while metro use rose 45 percent and suburban rail rose 139 percent). (Newman and Kenworthy 2015:90)

While Sao Paulo is not an automobile dependent city, it is most definitely is an automobile-saturated city which in particular is experiencing an escalation in the number of motorcycles which can better maneuver a megacity saturated with high-rise apartment buildings.

The culture of automobility has contributed profoundly to motor vehicle pollution and greenhouse gas emissions in many Latin American cities, including ones in Brazil, and Latin American countries in general. In Latin American countries and cities CO_2 emissions increased dramatically between

1980 and 2005 from road transportation, except for Cuba, a scenario upon which I touch in the next case study (Timilsina and Ashish Shrestha 2009).

CASE STUDY #4—KEEPING 1950s CARS ALIVE AND DISCOVERING ALTERNATIVE MODES OF TRANSPORTATION IN CUBA

Prior to the Cuban Revolution of 1959 middle-class Cubans, particularly in Havana, purchased US cars along with other US consumer items. In Havana, as Narotzky (2002:69) observes, "Chevrolets, Buicks, Fords, and Pontiacs were as ubiquitous as in any North American suburb: Havana in 1954 is said to have bought more Cadillacs than any other city in the world." In contrast to the United States where 1950s automobiles are generally regarded to be vintage collectors' items, Richard Schweid (2004:7) notes that in and around Havana, Studebakers, Nash Ramblers, Kaisers, Henry J's, Willys Overlands, De Sotos, Ford station wagons, and Chevrolet Bel Airs function as "working cars": "They are *ruteros*, taxis that run fixed routes beginning at the foot of the Capitol's steps and fanning out along the broad avenues and main thoroughfares of the city. All day long, crossing the chewed-up streets of Havana under a broiling sun in heavy traffic, they carry as many humans as can be squeezed in: three in the front seat and five behind is the minimum. The number of passengers who can be accommodated is often higher, as rudimentary seats are welded in place to fill any open space, and many of the cars from Detroit in the 1950s had plenty of open space." These aging but beloved cars for which spare parts have been unavailable from the United States since October 1960 when the embargo on trade with Cuba took effect and which formerly were owned by upper- and middle-class individuals number some 60,000, with most of them found in Santiago de Cuba and Havana. When affluent Cubans left their country, the revolutionary regime commandeered the automobiles and rewired and rebuilt them over the course of several decades.

Private taxis became *ruteros* under the auspices of the Servico de Transporte Popular (Schweid 2004:198). The Cuban government assigned refurbished American automobiles as well as imported cars from the Soviet Union and Eastern Europe to party elites, but only on an extended loan basis, perhaps anywhere from five to 15 years. In contrast to Fidel Castro who preferred to be chauffeured around in an Oldsmobile, Che Guevara generally preferred to drive his Chevrolet (Schweid 2004:199). Soviet Ladas and Volgas, East German Trabants and Czechoslovakian Skodas "were distributed to professionals—doctors and lawyers and university officials" (Schweid 2004:200). Some US models such as Fords and Chevrolets manufactured in Argentina slipped through the US-imposed embargo (Narotzky 2002:172).

During the Special Period beginning in 1990s when oil ceased to be imported from the Soviet Union, a shortage of gasoline resulted in a decreased use of not only cars but also buses and an increase in bicycle use. During the Special Period (1989–1993), Cuba's GDP decreased 35 percent due to fall of markets for its products and its exports fell 75 percent (Morgan et al. 2006). Severe food shortages resulted from a decline in food imports. Despite a decrease in daily energy intake from 2,899 calories to 1,863 calories per person, the shortage in fuel forced people to walk or cycle, translating into an increase in physically active adults from 30 percent to 67 percent with the average adult losing 9–11 pounds, or 5–6 percent of body weight (Murphy and Morgan 2013:333). Oil shortages forced a transportation revolution with trucks being converted into buses, the manufacture of local buses, and the extensive reliance on horse-drawn carriages and carts. The government implemented an extensive taxicab service. Despite its adoption of a Simpler Way in numerous respects, including in terms of transportation, in terms of the UN Human Development Index, Cuba ranks 52nd, the highest in Central America and the Caribbean, slightly below Argentina and Uruguay, and well ahead of India, ranked 134th, and China ranked 92nd.

CONCLUSION

This chapter has highlighted automobile utilization in four selected places, namely the two Germanys separated by the Cold War and now united with West Germany clearly politically and economically dominant over East Germany, Australia, Brazil, and Cuba. Obviously the cross-national and cross-cultural study of automobility could be extended to include many other regions, countries, and locales, as in part has already been the case. In contrast, Cuba has adapted itself to managing with less access to oil than it had prior to the collapse of the Soviet Union. Under the guise of the Bolivarian Revolution inspired by the late Hugo Chavez, Caracas with its easy access to petrodollars, which helped to lift millions of Venezuelans out of poverty due to a variety of social programs, is a megacity teeming with motor vehicles, including "muscle cars" from the 1970s and Jeep four-wheel drives from the 1990s (McGuirk 2014:139–141). Motor vehicles and roads constitute powerful symbols of modernity and nation-building in many developing countries, ranging from Albania to Peru and Mongolia (Pedersen, Axel, and Bunkenborg 2012; Dalakoglou 2012; Harvey and Knox 2012). Given the centrality of the automobile and other motor vehicles to modern life around the world, it is somewhat surprising that anthropologists, sociologists, and other social scientists have not given this domain more attention.

Africa constitutes a region that has been the focus of considerable anthropological research, on the part of both physical and sociocultural anthropologists. For particularly paleoanthropologists Africa has been an interesting place because it is the birthplace of humanity, whether we are speaking of the australopithecines as early hominines or *Homo sapiens* as the most recent hominine. Africa has been a focal point for European and North American anthropologists and is a glaring example that prompted Kathleen Gough (1968) to have dubbed anthropology as the child of imperialism. Despite a long tradition of anthropological research on Africa, it continues to be a last frontier of sorts in terms of mobility studies. Given the paucity of railroads in sub-Saharan Africa, motor vehicles have become the predominant mode of transportation for moving both people and goods around the region. Initially colonial powers created railway systems in their various colonies, but these were generally spotty in services both for carrying passengers and freight. Porter (2014:26) reports: "The expansion of motorized transport services followed rapidly after road construction in many parts of the continent from the early decades. Typically, they focused on linking major export producing areas to coastal ports along characteristically dendritic routeways, which were constrained within the boundaries of the colonial administrative power concerned. Trucks commonly provided feeder services to the railways, a key transport mode for the evacuation of agricultural produce and minerals for export in colonial Africa."

In the case of Nairobi, the Overseas Transport Motor Transport Company of London obtained monopoly rights in 1934 to operate a bus system called the Kenya Bus Services (Behrens, McCormick, Orero, and Ommeh 2017: 81). Similar bus services appeared in other sub-Saharan cities.

Following World War II and decolonization, railway systems in Africa went into decline and motor vehicles of different sorts, particularly trucks, buses, motorcycles and bikes, but also private automobiles and bicycles gained ascendancy in sub-Saharan Africa. The railway system in West Africa has contracted, except for Cameroon and some lines servicing private mining towns (Josa and Magrinya 2018:718). The World Bank has recommended that principal road networks be asphalted to stimulate trade and contribute to economic growth in various countries of sub-Saharan Africa (Josa and Magrinya 2018:724).

Unfortunately, in most sub-Saharan African cities extensive scheduled bus services have gone into decline. In their stead paratransit services relying upon mini- and midi-buses have appeared in cities throughout much of the region. Following Kenyan independence in 1963, state authorities permitted intercity *Matatus* or paratransit cooperatives to operate, initially illegally because they filled a gap in the official bus system but later legally (Behrens,

McCormick, Orero, and Ommeh 2017). Motorcycle-taxis are popular in both urban and rural areas. As more and more people in sub-Saharan Africa become relatively affluent, it is inevitable that automobile companies will target them more and more as potential customers.

The motor vehicle industry views sub-Saharan Africa as an open territory for expanding its markets and hopes to particularly sell its products to an emerging middle-class on the continent. The motor vehicle industry in sub-Saharan Africa has an open territory for expanding its markets and hopes to particularly sell its products to an emerging middle-class on the continent. In June 2018 Volkswagen opened a factory in Kigali, Rwanda, and already has a plant in Nigeria (Raji 2018:52). Mercedes-Benz plans to expand its East London plant in South Africa and India's Mahindra & Mahinda opened an assembly operation in South Africa in 2018 and plans to export automobiles to other African countries. Puegot Citroen plans to create an automobile assembly plant in Nigeria.

Toyota, the major motor vehicle company in Africa, sold 237,000 vehicles in the region in 2012 and has operations in all 54 countries, having most recently entered South Sudan. General Motors sold 167,000 motor vehicles in Africa in 2013, its primary markets being Egypt and South Africa. Ford, Nissan, Volkswagen, BMW, and Volvo also are selling motor vehicles in Africa. India's Tata and China's Foton companies have entered the African market, selling both automobiles and light trucks. Given the growing number of millionaires in sub-Saharan Africa, particularly South Africa and Nigeria, Porsche and BMW are targeting their luxury automobiles for this emerging market.

South Africa has emerged as the leading center of motor vehicles and component manufacturing in sub-Saharan Africa, accounting for 21 percent of the country's manufacturing out and 7.53 percent of its GDP (Damoense-Azevedo and Jordann 2011:156). The South African motor vehicle market underwent rapid growth from 1950 to the early 1980s but stagnated due to political developments revolving around apartheid and the early stages of the post-apartheid era. Since the beginning of the twenty-first century, production and sales of motor vehicles has expanded in South Africa. For example, in 2005 some 525,000 motor vehicles were produced in South Africa, with 26.6 percent of them having been exported (Black 2009:489).

Simba, a Kenyan company, presently assembles trucks, Chinese, and Indian automobile manufacturers (Raji 2018:51). Thomas (2018:15–16) reports: "[W]ith its reputation for cultivating skill jobs, high-value exports and local supply chains, the auto manufacturing sector appears to have a strong hand in negotiations with African governments, many of whom are willing to offer generous terms to secure the industry's attention. Ethiopia, Kenya and

Nigeria have all tinkered with a range of incentives in a bid to attract auto manufacturers, including tax breaks, tariffs on imports and the establishment of special economic zones." However, the South African automobile market clearly overshadows other sub-Saharan African markets, illustrated by the fact that in 2015, 6.1 million new cars were sold in South Africa whereas in Nigeria only some 300,000 cars were sold (MarketLine 2016:8).

Motor vehicle travel in sub-Saharan Africa often occurs by means of paratransit and motorcycles rather than conventional private automobiles. In the case of Kenya, the "'matatu' minibuses and midi-buses in Nairobi are reported to have the highest per capita use of informal transport in the world with 662 trips per inhabitant per year, three quarters of public transport trips and 36% of traffic volumes" (United Nations 2016:5). A decline in public bus systems in various sub-Saharan cities, such as Kampala in Uganda and Lagos in Nigeria, has contributed to the rise of alternative modes of transportation, including minibuses, shared taxi/vans, and motorcycles (Kumar 2011). Like other sub-Saharan African countries, which now have cities congested with a wide array of motor vehicles, the growing number of motor vehicles on the road in Accra and other parts of Ghana exemplifies growing disparities in the country. Hart observes (2016:184) that social inequality plays itself out "on the road, pitting the elite 'myself' drivers with luxury SUVs and sedans against working-class urban residents crammed in dilapidated trotos [a makeshift minivan], all asserting the rights to space, mobility, and opportunity in contemporary Ghana."

At any rate, from anthropological, social scientific, and historical perspectives, what are needed are more case studies of private motor vehicles in national and regional settings, particularly in places such as China, India, and other developing countries. Such studies of motor vehicles could touch upon the manufacture of motor vehicles in specific countries; what is the nexus between the motor vehicle industry and the state in various countries; who drives various types of motor vehicles; how is the culture of automobility shaped by the national culture; how have motor vehicles impacted upon settlement patterns and public transportation, the environment, and health in specific countries; etc. While some studies have ventured into these areas, there is still much research that needs to be done.

Chapter Seven

Beyond the Automobile

The automobile along with other motor vehicles has become a prominent commodity in virtually all societies in the world and increasingly particularly in developing societies. Are automobiles here to stay while they play havoc on settlement patterns, the natural environment, and health or is humanity at a historical stage at which they have become so dysfunctional that they must be transcended. Much has been written about new forms of automobility that will purportedly entail better designed and more environmentally sustainable automobiles, such as ultralight cars, electric cars, narrow-lane vehicles, and sub-cars, which would include "extremely lightweight two-, three- and four-wheel vehicles designed primarily for local and neighbourhood use" (Riley 1994:55). While motor vehicles over the past 40 years or so have become more energy efficient, at least in North America, Western Europe, and Australasia, this technological advance has been offset by increases in vehicle weight, power, and decreasing occupancy, resulting in constant historical energy intensities (Schaefer et al. 2009). Electric and hybrid automobiles potentially contribute less to both air pollution and greenhouse gas emissions, but not necessarily if they ultimately derive their electricity from coal-fired power plants. Even if they were to draw their electricity from massive solar power plants, they still would contribute to congestion in urban areas.

Automotive engineers have proposed the utilization of alternative fuels, including methanol, ethanol, natural gas, electricity, hydrogen, and reformatted gasoline (Riley 1994:157–266). Proposed new forms of automobility include "virtual automobility" in which "autonomous" or "smart cars" will operate on "smart roads" in which the driver can essentially function as a passenger, thus freeing him or her to engage in other activities (Miller 2006). Various cities in Belgium, France, Italy, and the UK are planning transportation systems that will accommodate autonomous cars. However, some or many drivers for a variety of reasons may resist the movement toward autonomous cars, even if the technology for them becomes perfected. Conversely, other drivers would embrace them with great enthusiasm. While there is no doubt that automobiles and other types of motor vehicles could be made to be more energy efficient, safer, less-polluting, and result in fewer greenhouse gas emissions, and so forth, all types of motor vehicles will still require massive amounts of natural resources, damage both the environment and individuals, and require massive swathes of land and space. While indeed motor vehicles on a limited basis will be required in remote areas, emergencies, and even deliveries, an increasing number of scholars and social activists are advocating a world beyond automobiles.

Despite the existence of massive corporate support for the ongoing use of motor vehicles, there have been some significant counterhegemonic efforts to resist the automobilization of society by emphasizing the need for people to rely on other forms of transportation. Referring to a society with the greatest number of cars in the world, Bohren (2009:377) correctly argues that the car is a "symbol of the American ethos of individualism, personal freedom, and mobility, shaped twentieth-century America." In recognizing the maladaptive nature of the car in the twenty-first century, she asserts that "[c]hange is needed in the human sociocultural practices that have accompanied the adaptation to the car in order to reduce its contribution to global climate change" (Bohren 2009:378). In his science fiction novel *Ecotopia*, Ernest Callenbach (1975:35) describes a fictional place situated in northern California, Oregon, and Washington State that has transcended cars. Aside from the question of whether such as place could exist in the modern world, as Peter Newman (2009:108) observes, "The biggest challenge in an age of radical resource-efficiency requirements will be a way to build fast rail systems for the scattered car-dependent cities." Anthropologist Catherine Lutz (2014:234) maintains altering the car system is a key element to solving a wide array of problems, including climate change; urban sprawl and other land use patterns that increase driving time, risk, and gasoline consumption; public health problems induced by automobiles such as fatalities, elevated levels of obesity, asthma, and cancer; and the reduced economic viability of auto-congested communities.

MODES OF CRITIQUES OF AUTOMOBILES

Perhaps the earliest and most strident of the critics of what he termed the "cult of the motorcar" was the renowned historian and literary critic Lewis Mumford, who wrote:

The current policy of wiping out every other means of transportation than the motorcar and the airplane has been fatal to both to the habitability of cities and to economy and efficiency in transportation. . . . [E]very metropolis that has encouraged the wholesale invasion of the automobile can bear testimony to its disintegrating effects, expanding every year with the expansion of the motorcar and petroleum industries: stalled traffic, personal confusion and frustration, excessive noise, poisoned air, increased costs in transporting goods, increased waste of time in fighting traffic jams and finding a place to park, increased necessity, as population expands at random, to spend an undue proportion of the day in performing scattered functions that were once more adequately carried on with the aid of wheeled vehicles in a compact urban area. (Mumford 1964:9–10)

In many ways, he anticipated later critiques of the culture of automobility. Matthew Paterson (2007: 36–58) delineates seven modes of critiques of automobiles which revolve around the following themes: (1) technocratic environmentalism; (2) safety, (3) automobile space: continued road building, (4) unequal movements, (5) atomistic individualism, (6) against speed consumerist, and (7) consumerist geopolitics. Much of the literature on automobility comes to terms with it at some level, implying that the automobile is here to stay. Conversely, at least portions of the "literature has moved from mere critique of the automobile and its effects to the claim that *automobility cannot continue*" (Reese 2016:154).

REFORMIST AND SYSTEM-CHALLENGING REFORMS IN ADDRESSING THE EXPANSION OF PRIVATE MOTOR VEHICLES

A plethora of suggested reforms have been made proposing how to make transportation, including automobiles, more environmentally sustainable. Despite the best of intentions, reforms are often problematic in that they may serve to stabilize capitalism as has been the case around the world repeatedly. Given this reality, Andre Gorz (1973) differentiates between *reformist reforms* and *non-reformist reforms*. He uses the term reformist reform to designate the conscious implementation of minor material improvements that avoid any alteration of the basic structure in the existing social system. Between the poles of reformist reform and complete structural transformation, Gorz identifies a category of applied work that he labels non-reformist reform or what I term a system-challenging reform. Here he refers to efforts aimed at making permanent changes in the social alignment of power. The distinction between these two types of reforms is sometimes hard to distinguish but one way might be whether they are initiated by the powers-that-be or whether they are initiated by the working class or various other subaltern groups or anti-systemic social movements.

Gorz (1973:5) maintains that the "ideal revolutionary change" would entail doing "away with the car in favour of the bicycle, the streetcar, the bus, and the driverless taxi," although many large commuter cities such as Los Angeles, Detroit, Houston, and even Brussels rule out this possibility given that they have been "built by and for the automobile." Challenges to the hegemony of the private car, both collective and individual, constitute examples of "non-reformist reforms" which challenge existing power relations and pave the way for more revolutionary changes in the larger society necessary for a more socially just and environmentally sustainable world. An example of a collective non-reformist reform would be the expansion of existing urban public transit systems and intercity passenger train systems and an example of an individual non-reformist reform would be restricting or even eliminating one's driving. Proposals made by various scholars, environmental and transportation associations, and other parties that seek to address the proliferation of motor vehicles often contain elements of both reformist reforms and non-reformist reforms. It can be difficult to determine which one of these designations is the most appropriate for a specific proposed reform.

Mumford (1964) made the following recommendations for containing the impact of the culture of automobility in Europe and the United States, particularly in the aftermath of World War II:

- The restoration and improvement of public transportation within cities
- Replanning neighborhoods to permit pedestrian flow and restriction of automobiles to them
- Developing small electric cars and restrictions on the presence of large automobiles in the CBD
- Relocating industry, business, and administrative operations to suburban sub-centers, resulting in cross-town traffic to the CBD

He believed that railways in their different forms—regular trains, subways, and electric trollies or trams, have a "spinal part of play in a balanced transportation system" (Mumford 1964:11). Mumford's approach to automobiles can be characterized as reformist rather than radical in that he did not advocate their total or drastic elimination.

Much more so than Mumford, the World Health Organization has been engaging in a reformist reform strategy on dealing with the issues surrounding road safety. In 2010 the United Nations General Assembly adopted a resolution that resulted in the creation of the Decade of Action for Road Safety (2011–2020), which encourages member states to make their roads safer. This resolution designated the WHO as a monitoring agency through its *Global Status Report on Road Safety* series, which has resulted in three reports to date. The WHO advocates the following measures in promoting road safety:

- Reducing speed
- Increasing motorcycle helmet use
- Reducing drunk-driving
- Increasing seat-belt use
- Increasing child restraint use in motor vehicles
- Reducing drug-driving
- Reducing distracted driving, such as occurs in the use of mobile phones
- Making motor vehicles and roads safer (World Health Organization 2015:21–56)

While all these measures or goals are in and of themselves laudable and worthwhile, nowhere in its report does the WHO advocate transcending motor vehicles, particularly automobiles and motorcycles, by encouraging UN member states to shift as much as possible to public transportation and to promote walking and cycling, the latter two which would go a long way to promoting health.

German sociologist Werner Broeg developed the TravelSmart scheme that is based on the "belief change toward less car dependence can work if it is community based and household," meaning that it has the potential to constitute a non-reformist reform, probably if it is coupled with public transport associations (Newman, Beatley, and Boyer 2017: 84). Travel Smart initiatives have emerged in Europe, particularly the UK, the United States, and Australia, including Brisbane and Perth. TravelSmart has introduced the "walking school concept" to empower children to safely walk to school. Newman, Beatley, and Boyer (2017:85) argue: "The experience of walking or cycling teaches children about their neighborhood and environment, as well as teaching road safety skills and equipping them for independence as they get older."

A common reformist reform when it comes to automobiles is to improve their fuel efficiency, namely the amount of fuel they consume per kilometer or mile. Schipper (2011:370) asserts that "[t]echnology has reduced the fuel required for a given car horsepower and weight markedly, but in the US until 2003 (and to some extent in Europe) this has been offset by greater new car power and weight)." This phenomenon is yet one more illustration of the Jevons Paradox or rebound effect by which improvements in energy efficiency are often followed by increased consumption of resources, particularly under a capitalist political economy which encourages continual economic growth.

Some large cities, such as London, Stockholm, Milan, and Singapore, have imposed a congestion charge which entails requiring drivers who enter the CBD to pay a fee. However, ultimately congestion charging constitutes, while perhaps reducing congestion in the CBD, yet another reformist-reform in that it allows affluent people to continue with business as usual. One strategy to theoretically compensate lower-income drivers is to provide them with discounts on registrations. A more radical approach to deal with congestion in the CBD is to restrict access to it for emergency vehicles, delivery vehicles, and vehicles carrying disabled people.

Another reformist reform is what goes under the designation of "eco-driving" which entails instructing drivers how to eliminate "bad driving habits that increase the amount of fuel they use for setting driving tasks" (Centre for International Economics 2016:42). While "eco-driving" may result in lower expenditures for the driver and slightly less pollution and greenhouse gas emissions, it preserves automobile dependence.

An increasing number of scholars over the course of the past two decades or so have made suggestions for reducing reliance upon automobiles or at least mitigating their worst impacts. Largely this is easier said than done. Bear in mind that over fifty years ago, Marshall McLuhan (1964:218) indicated that the automobile "will go the way of the horse" and would be replaced by other forms of transportation in a decade or so. As an integral component of capitalism, private motor vehicles have a great deal of staying power. Nevertheless, this grim reality should not deter us from developing ways to transcend them as much as possible.

Wright (1992) proposes the following steps: (1) switching from private motor vehicles to trains and buses; (2) increasing the distance that transport vehicles can travel per energy unit; (3) manufacturing engines that are less polluting; (4) implementing road designs, traffic regulations, and vehicle operations that contribute to more efficient vehicle utilization; and (5) relying upon less polluting sources of fuel. Newman and Kenworthy (1999:144–189) propose five policies for overcoming automobility dependency: (1) traffic calming in which speed plateaus, neck-downs, and other strategies are employed to slow down traffic in order to make streets safer, particularly for pedestrians, cyclists, shoppers, and residents; (2) the construction of quality transit systems as well as bike and walking paths; (3) the development of "urban villages" or multimodal centers with mixed, dense land use; (4) growth management to counter the urban sprawl; and (5) increasing taxes on motor vehicle transportation.

In addition to having a detrimental impact on the environment, cars are very expensive modes of transportation. Newman and Jennings (2008:45) report, "Cities that are car dependent spend between 15 and 20 percent of their wealth just on getting around, whereas transit-oriented cities spend only 5 to 8 percent of the wealth on transport." Consequently, Register (2001) maintains that cities should be designed for people, not cars. Furthermore, he proposes the notion of *pedestrian cities* in which people will not need cars and will be able to walk, cycle, or take public transportation to get around. In keeping with this proposal, Groningen, a city of some 170,000 in the Netherlands, removed the roads in its central business district in 1992 and adopted various policies that promote cycling (Korten 2001:256). Unfortunately, most of the efforts to make cities greener have benefitted the affluent much more than the poor and working-class people.

Sometime ago, Wolfgang Zuckerman (1991:74–157) provided a litany of "solutions" that could mitigate the worse features of automobility or even transcend them:

- Radical design of the conventional automobile
- Informing motor vehicle owners of the actual costs of their vehicles
- Restrict or abolish automobile advertising
- Impose total, partial, or temporary bans on the use of automobiles in inner cities
- Implement traffic calming measures and street modifications allowing pedestrians and cyclists a great share of street space
- Create mixed-use zones in which public transit vehicles, particularly buses and trams, allow pedestrians easy access to the zones
- Implement parking policies that restrict automobile use in cities
- Create policies that prevent or punish traffic violations effectively
- Impose strict regulation and higher taxation of trucks
- Make cities, suburbs, small towns, and rural areas more pedestrian friendly
- Encourage cycling by making it more appealing and creating a bicycle-friendly infrastructure
- Close off damaged roads to motor vehicles and open them up to pedestrians and cyclists

City governments in megacities in developed countries have adopted a number of strategies to restrict private motor vehicles utilization, including municipal value added taxes (VAT), tolls, congestion charges, and high parking fees. City governments in developing countries are more like to adopt automobile restriction. London imposed restrictions on automobile use with significant results and considerable revenue, but such a policy tends to favor the affluent who can afford to pay congestion charges (Van Themsche 2016:239).

Singapore has sought to limit the number of private vehicles within its boundaries, largely due to population growth and limited land, by implementing electronic road pricing and a policy target for 25 percent of all peak-hour travel to make use of public transportation by 2030 (Davison and Ping 2016: 213). Mexico City's mayor city created the "Hoy No Circulat" ("today it doesn't circulate") program in 1989 which restricts most drivers from using their automobiles one day a week, based on the last digit of the car's license plate, to reduce congestion and air pollution. Bogota, Santiago, and Sao Paulo emulated this program, but they ultimately failed. In the case of Mexico City, more affluent residents purchased a second automobile "with a license plate that didn't match the same day or used taxis" (Van Themsche 2016: 239). The second automobile that residents purchased often was a second-hand one which created more pollution than their primary automobile. Fortunately, Mexico City has made some headway in developing a suburban train system.

Following Freund and Martin (1993:60), a political ecological approach to addressing the health consequences of automobility requires "changing the social and physical environment (e.g., building safer highways), producing safer cars, and making many alternative ways of traveling available to drivers." However, even the construction of safer highways and production of safer cars will not be enough because the health benefits will be far greater in moving away from an infrastructure that perpetuates motor vehicle utilization. In this transition, cities would have to be redesigned to accommodate emergency and delivery vehicles and special vehicles to transport disabled people. Furthermore, the development of green cities would have to be part and parcel of achieving social equality and hopefully will give an increasing number of people options in their choices of where to live.

John Urry maintains that "many significant cracks in the car system" have become more apparent over the past two decades:

- Policy makers, transport planners, and the automobile industry have become more aware of the health, environmental and energy problems posed by the automobile, "which seem impossible to "solve" in any simple sense" (Urry 2016:134)
- A commitment to policy makers and motor vehicle manufacturers to "business-as-usual" approach to the car system is disintegrating
- "there is much experimentation by small and medium-sized enterprises, NGOs, city governments and large corporations in developing alternative charging systems, body shapes, batteries, fuel types, and much lighter body materials" (Urry 2016:134)
- "urban design is being developed towards car restraint, including parking restrictions and tariffs, traffic calming, pedestrianized city centres, bus

lanes, bicycle tracks, public bike schemes, road pricing and so on" (Urry 2016:134)
- Growing evidence indicates that cities with less automobile dependence and more city center living exhibit higher levels of well-being
- Less automobile-dependency, including among young people, appears to be occurring in developed societies
- "there is evidence from some countries that younger generations prefer smartphones to cars, if given a choice, with two-thirds of young people stating they like spending money on 'new technology' rather than cars" (Urry 2016:135)

Despite such cracks in the culture of automobility, particularly in developed societies, significant technological, political, economic, and sociocultural barriers continue to exist in drastically revamping the present system, perhaps particularly in developing societies which have only started to embrace private automobiles in significant numbers in recent decades. Various scholars, perhaps inspired by the literature on "peak oil," maintain that at least developed societies may be reaching an era of the "peak car" (Goodwin and van Dender 2013). Metz (2013:259) observes that in the case of the UK "car and van ownership has stabilised at 75% of households, probably reflecting substantial saturation of demand on the part of those able and wishing to drive (currently 72% of adults in Britain hold a driving license)." While London has increased in population over the past two decades or so, automobile trips have declined from 50 percent of all journeys in 1993 to 38 percent in 2013. At the same time, reliance on public transportation has increased in London. Urbanization has contributed to a decline in automobile use in several other developed societies. While there was an increase in motorization rates in five Central European cities, namely Berlin, Zurich, Vienna, Hamburg, and Munich between 1970 and 2000, there was a slight decrease in the motorization rate between 2000 and 2010 or 2007. Conversely, many European and Asian cities have begun to develop suburbs outside of their existing transit zone, contributing to automobile-dependency on the part of their residents (Newman and Kenworthy 2015:110).

The US Census reports a decline in the percentage of automobile commuters from 83 percent in 2000 to 86 percent in 2014 (US Census Bureau, 2000–2015). Some US cities underwent larger reduction during this period, such as Portland, Oregon, from 76 to 67 percent; Minneapolis from 73 to 68 percent; Washington, DC, from 33 to 27 percent: and New York City from 33 to 27 percent. Conversely, outer suburban and rural towns and areas in developed societies tend to be highly car-dependent, a pattern that I have witnessed first-hand in my travels around Australia. Evidence from various developed societies and regions, including North America, UK, Western Europe, Japan, and Australia, indicate a growing percentage of young people are not obtaining a driver's license, and even if they have one, are less likely to drive. These trends have been modest, varying from declines of 0.4 to 1.2 percent over the past decade or two in countries such as Australia, the United States, Norway, Sweden, the UK, Japan, Germany, France, and Canada, but at the same time there have been increases between 0.3 and 1.2 percent in young people obtaining licenses in Finland, Israel, and the Netherlands (Delbosc and Currie 2013:274). Based upon a survey of the literature, Delbosc and Currie (2013:277–278) delineate the following reasons for why young people in at least certain developed countries are less likely to obtain a driver's license:

- Life stage
  - Increased rate of formal studies
  - Reduced job opportunities
  - Delayed marriage and child-bearing
  - Living longer with parents
- Affordability
  - High cost of automobile insurance
  - High cost of gasoline or petrol
  - High cost of automobiles
  - Economic downturns
- Location and transportation
  - Reliance on public transportation or cycling or walking
  - Relocation in inner-city areas
- Driver licensing regulations
- Attitudes
  - Environmental consciousness
  - Decline of automobile as status symbol
  - Involvement in other activities
- E-communication

Klein and Smart (2017: 20) drawing upon data from the Panel Study of Income Dynamics assert that while American Millennials overall own fewer automobiles compared to previous generations, when they controlled for "whether young adults have become economically independent from their parents, i.e. left the nest, . . . the economically independent young adults own cars." Reporting on the situation in the UK, there is evidence that the overall use of automobiles may be stabilizing for the time being in so-called Western economies: "Since the late 1980s the proportion of people of any one age with full car access has changed significantly. . . . Between 1988 and

1995 around 70% of men had full car access by the age of 30, and the figure for women was around 45%. However, by 2008–10 the rates were broadly similar at around 55% for each and the profiles of growth up to the age of 40 virtually indistinguishable" (Stokes 2013:365).

Evidence from Munich, Berlin, Vienna, and Zurich indicate that there has been a significant reduction in the automobile share of trips over the past 25 years, even though many people continue to own automobiles (Buechler et al. 2017). Conversely, these cities have implemented programs that make automobile travel more expensive and slower and that promote public transit, walking, and cycling. Public planning on numerous social and environmental issues is generally more acceptable in Europe than it is in the United States and even Australia.

Based upon an analysis of National Household Travel Survey data between 2001 and 2009 Brown et al. (2016:63) assert that the higher use of public transit among young people, particularly those of color, in the United States may not persist as they become older unless "concerted efforts to motivate increased transit use and improve transit attractiveness" are implemented. Newman and Kenworthy (2015:73) maintain that while there appears to be wide-spread decline of automobile dependence in developed societies, they admit "there are still a huge number of cars on the streets of the world's cities, and in many cases this number is still growing."

Some developed cities have opted to close freeways. Following the collapse of a section of the West Side Highway in 1973, most of the throughway was shut down, with no apparent sign of traffic congestion in adjacent areas (Newman and Kenworthy 2015:152). Following the San Francisco earthquake of 1989 which destroyed the Embarcadero Freeway, authorities opted not to rebuild this highway, resulting in a revitalization of the city's waterfront. While the overall trend toward a decline in automobile use in many developed societies is a positive development, it is being off-set by the increased ownership of private motor vehicles, not only automobiles but also motorcycles and motor bikes, in developing societies. Furthermore, it is possible that as young people in developed societies complete their education, obtain steady employment, marry and have children, and move to the outer suburbs they will purchase and drive automobiles.

Martin and Shaheen (2010) conducted a survey of 6,895 US residents and 2,740 residents about the extent of their driving and how it impacts on the generation of greenhouse gas emissions. Overall, they conclude that the "average change in emissions across all respondents is 0.58 tGHG per household per year for the observed impact, and 0.84 tGHG per household per year for the full impact" (Martin and Shaheen 2010:3). The "observed impact" refers to the emission change that occurred whereas the "full impact" includes the observed impact but also avoided emissions. Individuals who formerly owned cars tend to drive less when they join a car sharing scheme but individuals who had been carless inevitably drive more when they join a car sharing scheme.

Car Sharing

Martin and Shaheen (2010:4) report: "[T]he results show that a majority of households are increasing their emissions through carsharing—but the degree to which these households are increasing their emissions is very small. In contrast, the minority of households reducing their emissions are exhibiting changes that are of larger magnitude and greater variance" (Martin and Shaheen 2010:4). This study suggests that while car sharing in terms of serving as a climate change mitigation strategy is at best very modest and clearly constitutes a reformist reform.

Car sharing diffused to Sydney in 2003 and by 2013, the number of shared vehicles had increased to over a thousand, serving some 20,000 members, a modest number given that the city population of the city in 2011 was about 4.4 million people (Dowling and Kent 2015: 58). Travis Kalanick, the founder and CEO of Uber, has boasted that his company will "make car ownership a thing of the past" (quoted in Hill 2015:3). Indeed, Uber has expanded into some 300 cities and 58 countries world-wide and has posed an economic threat to standard taxi companies. The company refers to its drivers as "partners," but 80 percent of Uber drivers work part-time and by and large accrue relatively meager earnings. Uber's operations have been either shut down or partially restricted in Virginia, Maryland, South Carolina, Nevada, New York City, Miami, Philadelphia, Chicago, Birmingham, New Orleans, San Antonio, and many other places for a number of reasons, including city ordinances and pressure from taxi cab drivers and their companies (Hill 2015:85). Uber also has encountered competition from other ride share companies, particularly Lyft, which reportedly treats its drivers better.

Lane et al. (2015) published a World Resource Institute report calling for car sharing as one strategy for addressing the growing proliferation of private automobiles in developing countries or what they term "emerging markets." Ironically their study received funding from the Shell Foundation, Stephen M. Ross Philanthropies, and the Volkswagen Research Group. In the foreword to the report, Andrew Steer, the President of the CEO of the World Institute, makes the following pleas for car sharing as strategy for addressing problems of urban sprawl, pollution, and health stemming from the growing number of motor vehicles in the cities of developing countries:

- Individuals obtain the benefits of automobile use without the high costs of ownership
- Car sharing has the "potential to increase carless households' mobility and access to goods, services and opportunities and, in some markets, it may even delay the decision to purchase a car"
- Car sharers might only occasionally use an automobile and move about by other means of mobility, such as walking, cycling, and public transit (see Lane et al. 2015:1)

While as is the case in developed societies (see Millard-Ball 2005 et al.), car sharing in developing societies would constitute a preferable option or the "lesser of two evils" to outright automobile ownership. Low, Gleeson, Green, and Radovic (2005:158) report: "Car-sharing schemes in Switzerland and Germany have halved the distance travelled by car annually by participants in the scheme. Liftshare in the UK uses the internet to register people offering and needing transport from place to place."

Ultimately, however, car sharing clearly still constitutes a reformist reform in terms of challenging the culture of automobility. Not surprisingly, the car sharing industry's "target demographics in emerging markets seems to have similar demographics in mature markets: well-educated, mostly carless, middle-income, young to middle-aged, urban residents" (Lane 2015:4). The researchers recognize that the introduction of car sharing in developing societies faces tremendous barriers, including traffic congestion, competition from auto-rickshaws and taxis, and limited street parking space. They do not consider that car sharing may only serve as a stepping stone to full car ownership once an upwardly mobile person obtains the financial means to purchase an automobile of his or her own. In a truly sharing economy, namely an eco-socialist one that I discuss in the concluding chapter, a car-sharing system in which private automobiles have vanished and been replaced by publicly owned automobiles, car-sharing has the potential of vastly reducing their numbers. Harari (2015:384–385) convincingly argues that under a car-sharing system, "50 million communal autonomous cars may replace 1 billion private cars, and we would also need far fewer roads, bridges, tunnels, and parking spaces. Provided, of course, I renounced my privacy and allow the algorithms to always know where I am and where I want to go."

Because air pollution in Chinese cities has reached a critical level in terms of impact on health, as part of its 12th Five Year Plan the Chinese state implemented in 2005 the Transportation Demand Management scheme which "directly involves reducing the desirability to access a car, and a large-scale shift toward public transit opportunities" which reportedly has resulted in levelling the manufacture of motor vehicles to only 0.8 percent increase in 2011, following years of very high increases per annum (Newman and Kenworthy 2015:82). Urban transit and cycling are being prioritized in many Chinese cities. Furthermore, the government has fostered the development of high-speed trains, resulting in over 8,500 kilometers of high-speed rail track in the country (Newman and Kenworthy 2015:85). Many cities that had privatized their bus services socialized them into subsidized public-sector operations (Mehndiratta and Salzberg 2012).

Dennis and Urry (2009) have boldly asserted that humanity will shift from the present car system to a new "post-car system" over the course of the next few decades. Their post-car system consists of the following components (Dennis and Urry 2009:62–108).

- New fuel systems that might include biofuels, electric batteries, and hydrogen
- New materials replacing steel such as polymer composites, aluminium, nanotechnology, and carbon-based fibers
- Smart vehicles in which drivers and passengers communicate with other drivers and passengers so to facilitate a smooth flow of traffic
- Digitalization in which traffic is the "social sorting of travel through digitizing networks of road travel"
- De-privatizing vehicles in which people share cars on a scheduled basis
- New transport policies which encourage people to use travel modes other than private motor vehicles
- New living, work, and leisure practices which reduce the amount of traveling for these activities
- "Disruptive technologies that foster low carbon alternatives to existing forms of transportation"

Sustainable transportation would entail many other measures, such as limiting the use of cars as much as possible, making them smaller and more energy efficient, and even banning four-wheel drives except in special circumstances (such as in the outback and rugged mountainous areas) and drastically limiting air travel.

Electric Cars

Electric cars are often offered as a more environmentally sustainable form of transportation because they are more energy efficient. Engine efficiency refers to the "relationship between the total energy contained in fuel and the amount of energy used to propel cars" (Van Themsche 2016: 169). Compared to the internal combustion engine, electric motors are much more efficient in that "1 kWh of energy will propel a typical electric car 2.94 miles, compared to 0.83 miles for a similarly sized gasoline-powered vehicle" (Heinberg and Fridley 2016:84). Electric engines have a much higher energy efficiency, about 95 percent, as opposed to 25 percent energy efficiency for gasoline engines and 30 percent for diesel engines (Van Themsche 2016: 169). Coffman, Bernstein, and Wee (2017:79) observe: "EVs are a potentially important technology to help reduce greenhouse gas emissions, local air pollution, and vehicular noise. . . . In recognition of these benefits, countries around the world are setting EV adoption targets." This might be partially the case if they derive their power from renewable sources of energy but not necessarily if they derive their energy from coal-fired power plants. Electric vehicles sales reached one million worldwide in 2015, with about a third of sales being in the United States, another third in Europe, 17 percent in China, and eleven percent in Japan (Coffman, Bernstein, and Wee 2017: 79–80). The 50 most populous urban areas in the United States accounted for 81 percent of the 2016 US electric vehicle market, with a concentration of the market based on the West Coast, particularly San Jose leading the way (Slowik and Lutsey 2017). The driving range of electric automobiles varies greatly with the longest range being 424 kilometers for the 2014 Tesla Model S and the shortest range being 50 kilometers for the 2013 Scion iQ EV, with most electric automobiles having ranges between 100 and 160 kilometers (Huang, Kanaroglou, and Zhang 2016:1). Furthermore, electric cars will not solve congestion problems and the need to build and maintain roads, which requires an enormous amount of concrete, the manufacture of which produces CO_2 emissions. As Newman and Kenworthy (2015:183) observe, "Roads in automobile-dependent cities are already full. It makes no sense to imagine traffic increases based on electric vehicles if they carry on destroying the walking city and transit city fabrics, just as petrol and diesel vehicles have done." Furthermore, for the foreseeable future electric automobiles will be much more expensive than conventional automobiles and thus only available to affluent people. While the operating costs for electric cars is only about a third of that for fuel cars, the purchase and battery costs for the electric cars make them unaffordable for ordinary consumers. As Prud'homme and Koning (2012: 64) observe, other than a "handful of consumers who want to show they are very rich and very green, there is not much of a market for the electric car as it is presently." Furthermore, the manufacture of electric vehicle batteries requires large amounts of energy and rare minerals (McManus 2012). Electric automobiles and buses operate on batteries with a lithium requirement, a resource that exists in limited supply. Li (2016:119) reports: "If the entire world's annual lithium production (at the rate of 2013) is used to make batteries for electric vehicles, it will take about 230 years to replace the world's current car fleet with electric vehicles. This has not accounted for the growth of the world car fleet as a function of world economic growth nor the additional lithium consumption that will be needed to replace worn-out lithium batteries." Batteries for electric automobiles are heavy and need to be frequently recharged. In so many words, electric automobiles are not in the long run a viable alternative to conventional automobiles, at least under the parameters of existing technology. As Haas and Sander (2016:131) observe, ultimately "electric cars do not challenge the fundamental structures of the automotive society that wastes a lot of resources, plasters the entire country with a dense web of streets and motorways, makes cities a hostile place to live, and, not least, claims many people's lives due to air pollution and accidents." In a similar vein, Pirani (2018:186) asserts that while electric automobiles fit into a "green growth" framework, at least as green as their source of electricity, they "do nothing to reduce quantities of fossil fuels consumed in making roads, parking spaces, and cars themselves."

However, given that even in the best of all worlds there will be a need for delivery, emergency, and police vehicles, vans for highly disabled and elderly individuals as well as buses to connect with train and tram lines, electric vehicles powered by renewable energy sources will be needed as a replacement for conventional motor vehicles (Wikstroem, Hansson, and Alvfors 2016). Unfortunately, large trucks, the primary means for moving freight in the United States and many other countries, "require batteries too heavy to be practical in most instances, particularly if they are traveling long distances" (Heinberg and Fridley 2016:84). Conversely, some bus rapid transit (BRT) systems have adopted battery-driven vehicles and there appears to be potential for the development of electric intercity buses (Heinberg and Fridley 2016:85).

Autonomous or Driver-less Vehicles (AVs)

The latest rage in various circles is the driver-less, electric autonomous car, one which is monitored by computer algorithms which integrate them into a single network. It is commonly believed that self-driving or autonomous automobiles will be commercially available in a few years and that they will be more energy efficient and safer (Davis 2016: 179–181). S. Van Themsche, a staunch proponent of e-mobility, boldly asserts:

> Is e-mobility just another buzzword used by big corporations to sell a green image or is there behind the marketing appeal of this word, real fundamental technological breakthroughs? We believe that e-mobility could help save millions of passengers from injuries, reduce millions of deaths from pollution-related diseases, bring a higher level and better quality of life, reduce global warming,

and decrease transportation costs. This might seem too good to be true, but the reality is that under the convergence of IT and wireless communication, and associating the increase in power electronics together with the improvement in battery performance, fantastic new opportunities are being unleashed to reduce energy consumption and increase transport capacity, while improving the passenger's journey experience. (Van Themsche 2016: 2)

He believes that the creation of an increasing number of powerful algorithms is facilitating the e-mobility revolution (Van Themsche 2016: 19). Van Themsche (2016: 375) asserts that given that in many developed countries families own more than one car, access to an autonomous car would allow the family to own only one car since an "autonomous car would be able to drop one family member and return home to pick-up the other members." However, what happens when two or more members of the household decide that they want to leave for work or school or whatever about the same time, such as in the morning? Theoretically, autonomous cars would operate 24/7, other than during downtimes for maintenance, and would reduce the total number of cars on the planet, most of which sit idle 90–95 percent of the time, along with the need for parking space required by these idle cars (Van Themsche 2016: 380). Theoretically, autonomous automobiles, at least if their technology would be virtually fail-proof, could eliminate private ownership of vehicles and could be part and parcel of the moving to an entirely socialized transportation system.

Proponents of autonomous vehicles from various quarters propose three frameworks for their implication: (1) private ownership of AVs by individuals or households, (2) private ownership of AV fleets; and (3) public ownership of AV fleets by local governments, comparable to existing public transit systems (Hawkins and Habib 2019:67). Google announced in 2010 that it has been testing a fleet of autonomous vehicles and Tesla advertised the availability of full self-driving capability on its 2017 automobiles (Lutin 2018:92). Nissan plans to sell autonomous automobiles by 2020 and Tesla, Jaguar, Land Rover, Ford, and various other corporations predict that they will be selling autonomous automobiles in the next several years. However, Newman, Beatley, and Heather Boyer (2017: 64), unabashed public transportation advocates, assert that the "vehicle industry is trying desperately to hold onto market share when it should probably be looking at the decline and stranding of mobility assets due to peak car use [at least in developed societies], or reduced car use per person."

Self-driving pilot programs exist in Europe, the United States, and Australia. General Motors operates a large driverless automobile testing facility in the Detroit area and Toyota has one near Mount Fuji (Van Themsche 2016: 379). The German Ministry of Transport plans to convert part of the A9 freeway between Berlin and Munich into a test track for autonomous vehicles (Bunzhez 2015: 452). The EU CityMobil2 project has successfully demonstrated low-speed autonomous vehicles in seven European cities (Lutin 2018:93). Autonomous vehicles are becoming a part of the public transit planning process in various urban areas, including the United States where some cities have been conducting demonstrations of low-speed, 10–12 passenger AVs (Lutin 2017:94).

Aside of the embedded energy in such a system, the transition from the present one in which motor vehicles by and large have a driver to one in which they all lack a driver will be an awkward one to say the least as is illustrated in the following scenario described by Harari (2015): "In August 2015, one of Google's experimental self-driving cars had an accident. As it approached a crossing and detected pedestrians wishing to cross, it applied its brakes. A moment later it was hit from behind by a sedan whose careless human driver was perhaps contemplating the mysteries of the universe instead of the watching the road. This could not have happened if *both* vehicles were steered by interlinked computers." However, what he appears to overlook is that even the best designed driver-less car system may periodically have operating malfunctions which could result not only in isolated accidents such as the one described above but massive collisions on a busy freeway.

On a broader scale, Newman, Bentley, and Boyer (2017) express strong skepticism about the frequent claims that autonomous automobiles will prove to be safer than conventional automobiles. For example, they assert, aside of the removal of human intervention, a "troop of cars doing 100 kph with 1 meter spacing will constantly be interrupted by those wanting to take over control to get out of the column or through some preference or panic," thus creating the possibility of immense accidents (Newman, Bentley, and Boyer 2017: 62). Furthermore, when autonomous automobiles exit the freeway, "they will be in a much less predictable space shared with other vehicles, cyclists, and pedestrians that are not under any kind of control system." As Dixit, Chand, and Nair (2016) observe, "there are many safety and risk related unknowns associated with the autonomous vehicles, with regards to factors affecting disengagements and driver behaviour, at moments requiring manual resumption of vehicle control." Even if the technology behind driver-less cars or motor can be perfected, a possibility that should not be ruled out by any means, Harper et al. (2016: 1) argue that they "represent a technology that promises to increase mobility for many groups, including the senior population (those over age 65) but also for non-drivers and people with medical conditions." While mobility for those lacking it is badly needed, there are more environmentally sustainable methods for providing elderly and disabled people with mobility, such as shuttle vans, the availability of which should be increased and on a low or no cost basis. How autonomous vehicles interact with pedestrians and cyclists is an issue that has only begun to receive attention. Skippon and Read (2017:6) observe: "In the absence of human drivers, AVs may need to adopt new forms of communication to indicate their intent to pedestrians. Similarly, pedestrian behaviours may adapt to automated vehicles in the knowledge that, within the capabilities of its braking performance, an automated vehicle will certainly stop for a pedestrian in the roadway."

Ultimately, the notion of autonomous cars could be extended to other vehicles, particularly buses and trucks, as it already has been to trains in some places. Even more futuristic than the autonomous car is the concept of the flying car. Supposedly flying cars would "fly lower than commercial planes [albeit not necessarily helicopters] and thus wouldn't be controlled by visual flight rules traffic" (Van Themsche 2016: 368). They would function similarly to present-day drones Although Van Themsche (2016: 368), who appears to be as much an enthusiast about flying cars as he is about driverless cars, admits that modern "risk society" might resist the former because an accident might result not only in fatalities to the passengers but also to pedestrians, homeowners, or presumably passengers in driver-less cars below.

At any rate, unmanned technology, whether it is for automobiles, trucks, buses, or trains, will rely upon highly sophisticated technologies that "will need to include smart traffic management system, which will . . . still require traffic lights for pedestrians or cyclists, intelligent cameras to identify road blocking objects, and panel systems informing about changes in road conditions" (Van Themsche 2016: 389). Legacy et al. (2018:92) suggest that AVs "may reinforce existing automobility-based hegemonies whereby the future of mobility will see that the car will remain centre-stages and individually owned." Furthermore, some experienced motorists do not desire AVs, desiring a sense of control as well as the excitement and thrill that driving provides for them. Various studies suggest that AVs may contribute to increased urban sprawl in that they may also induce their passengers to travel more frequently and across greater distances (Hawkins and Habib 2019:69). Streets and highways may also become more congested due to the presence of empty AVs either sitting or moving around. While the motor vehicle maintains that AVs will free up passengers, no longer drivers as such, with time to engage in various productive and recreational activities, functioning essentially as mobile offices, sleeping chambers, and entertainment centers, Singleton (2019:54) asserts that claims of the "productive use of travel time in AVs may be smaller than proponents suppose."

Social Activists Opposing Cars and Highways and Promoting Public Transportation

Fortunately, challenges to the hegemony of the private motor vehicles are coming from many quarters, including environmentalists, public transportation advocates, radical cyclists belonging to Critical Mass who ride through busy streets to express their opposition to a car-oriented society, progressive politicians, and even academics. Environmentalists and other social activists began to challenge the pollution, health hazards, traffic congestion, urban sprawl, and fragmentation of social life resulting from motor vehicles and highways in the 1960s and 1970s, a period of social ferment on many fronts around the world (Golten et al. 1977). Ladd (2008:133) asserts: "Antifreeway activists joined lovers of city life, conservationists (soon to be much more numerous and known as environmentalists), urban politicians, and a growing number of transportation planners in promoting a revival of mass transit during the 1960s. The car, they believed, was reaching the limits of its usefulness, even in the suburbs."

In the United States, in large part due to campaigning on the part of various environmentalists, Congress passed the 1964 Wilderness Act which prohibited the creation of permanent roads as well as the use of motor vehicles, motorized equipment, motorboats, the landing of any aircraft, any form of mechanical transportation, and the erection of any structure or installation. Particularly prominent in the campaign lobbying for the passage of the Wilderness Act were Robert Marshall, Aldo Leopold, Robert Sterling Yard, and Benton MacKaye (Sutter 2002). Marshall played a key role in persuading the US Forest Service and the Bureau of Land Management that they should set aside wilderness areas on the lands they managed. Leopold was a prominent wildlife ecologist who pushed for wilderness protection during the 1930s and 1940s. Yard, who was involved in the early development of the National Park Service, concluded that the parks were inadequately protecting wilderness areas and served as a pioneer in the Wilderness Society. MacKaye was a regional planner who played a key role in the eventual establishment of the Appalachian Trail. Of these staunch wilderness advocates, Marshall and MacKaye exhibited socialist sympathies.

More recently, the SUV or sports utility vehicle or what in Australia is generally simply called a four-wheel drive "has become the lightning rod for critics of American patterns of consumption and inefficiency" (Vanderheiden 2010: 24). The anti-SUV movement includes groups such as the Friends of the Earth, Earth on Empty, and the Evangelical Environmental Movement. Various environmentalists view driving SUVs and Hummers as a manifestation of "foolish" consumption. "Fossil Fools Day" has become widespread on many university and college campuses as well as in communities throughout

North America, particularly in the Pacific Northwest. Supporters of "Fossil Fools Day" encourage people to overcome their addiction to oil by not driving, if at all possible, by walking, cycling, or relying on public transportation and, in extreme circumstances, car-pooling or driving fuel-efficient automobiles, shopping locally, and purchasing food items that are produced locally or regionally instead of from far-away places.

Pockets of resistance to motor vehicle hegemony are manifested in the slow but steady development of a "global auto city protest movement" (Newman and Kenworthy 1999:60–62). Many environmentalists view reliance on the automobiles as "irresponsible squandering of natural resources" and a major contributor to air pollution and climate change (Ladd 2008:140). Kay (1997:286) asserts "that deposing the car from its dominion over the earth is a radical, even revolutionary, move" and argues that those who participate in this still burgeoning "countercultural rescue movement" must act as "promobility advocates: pro-walking, pro-cycling, pro-transit" (Kay 1997:286). Kay (1997:356–357) advocates a strategy of anti-automobile activism at the local, regional, state, and national levels that challenges "moribund highway-based plans" and the "vehicle-first policies promoted by long-entrenched forces."

Grass-roots groups opposed to highway construction projects that threaten stable and historic neighborhoods and rural landscapes have formed in states such as Oregon, California, Kansas, Indiana, and New Hampshire. In the early 1950s residents opposed the demolition that was necessary for constructing New York's Cross-Bronx Expressway (Ladd 2008:109). San Francisco residents resisted proposals to rip up the city's iconic cable car system and managed to preserve three cable-car lines and some electric trolley lines (Ladd 2008:132). Protests opposing freeway development evolved into a national movement in the United States during the late 1960s and spread to other countries. Eric Avila (2014) maintains, however, that the anti-freeway campaigns found the greatest success in the United States tended to occur in white middle-class or affluent communities, such as Cambridge, Massachusetts; Lower Manhattan; the French Quarter in New Orleans; Georgetown in Washington DC; Beverly Hills, California; Princeton, New Jersey; and Fells Point in Baltimore. Fellman (1986:35) found that middle-class people played a pivotal role in opposition to a proposed freeway to run along Brookline and Elm Streets in Cambridge, Massachusetts, in the early 1960s for two reasons: "First, their professional training enables them to read and to use road design specifications, both to criticize official plans and to create counterplans. Second, the middle-class participants "speak the same language" as government officials, meeting them on their own terms as members of the same class."

Nevertheless, people of color in the United States have also mounted anti-freeway campaigns, as has been the case for Hispanics in Los Angeles, African Americans in Detroit, and native Hawaiians in the Honolulu metropolitan area. The Labor/Community Strategy Center in Los Angles has challenged the underfunding of the public bus system upon which poor workers, many of them people of color, rely (Martin 2002:17).

Grass-roots groups in both Toronto and Vancouver prevented freeways from being built into the inner city (Newman and Kenworthy 1999), and over 900 anti-freeway groups have emerged throughout Britain (Newman and Kenworthy 1999). While protests against the construction of the M4 between London and Bristol were muted and fragmented, opposition to this motorway came from groups such as the Ramblers' Association, the Farningdon branch of the National Farmers' Union, and the Berkshire branch of the Council for Preservation of Rural England (Mees 2010:61).

It is important to note that not all people who have been critical of mainstream society have been historically critical of the culture of automobility. As Foster (2003:136) observes: "The 'flower children' may have gathered in Haight Asbury in San Francisco for the 'summer of love' in 1967, but many of them arrived by car. In fact, one of the most cherished symbols of hippiedom was a beat-up Volkwagen bus painted in psychedelic rainbow colors and festooned with peace signs and radical bumper stickers."

Conversely, even in the highly automobile dependent United States, cycling is on the rise, not only for recreational purposes but also as a form of daily transportation or commuting. Furness (2010:3) reports: "New York and Chicago saw 77 percent and 80 percent increases in bicycle use between 2000 and 2008, while Portland, Oregon, a city boasting one of the highest rates of cyclists in the country as well as a vast cycling infrastructure and a vivid culture of bike devotees, witnessed a 144 percent increase in bicycle use between 2000 and 2008." Despite such upward trends in bicycle ridership, only about one percent of Americans cycle as a form of transportation. In contrast, however, 27 percent of trips made in the Netherlands are by bicycle, 18 percent in Denmark, and about 10 percent in Germany, Finland, and Sweden (Furness 2010:4). Nevertheless, some Americans view the bicycle as a "source of self-empowerment and pleasure, a pedagogical machine, a vehicle for community building, a symbol of resistance against the automobile and oil industries, and a tool for technological, spatial, and cultural critique" (Furness 2010:9). The bicycle historically has been portrayed as a liberator of women, an efficient use of natural resources, a means to challenge authority, an alternative form of mobility to the automobile. The countercultural movement in Europe and North America has played a pivotal role in the rise of modern bicycle advocacy. In reality, bicycle advocates today range from activists who engage in a radical critique of modern technology and capitalism to those "more in tune with the civic engagement of people like John Dewey and Ralph Nader" (Furness 2010:68).

Critical Mass, an organization of pro-cycling activists, has engaged in pro-bicycle and anti-automobile mass actions in cities such as San Francisco, Austin (Texas), Washington, DC, and Edmonton. It began in San Francisco in 1992 when bicycle commuters decided to converge at rush hour on the last Friday of every month with the message: "We are not blocking traffic, we *are* traffic." As Furness (2007:299–300) observes, "Critical Mass bicyclists use spontaneity, playfulness, and decentralized organization as ways to raise fundamental questions about the nature of automobility, the polemics of car culture, and the (mis)use of public space." Today small groups of cyclists in Europe and North America view the bicycle as the locus for a rising critique of automobiles and automobility (Furness 2010). Conversely, *Vehicularists* oppose the radical tactics of Critical Mass, arguing that "bicyclists fare best when they act, and are treated in return, as drivers of vehicles" (quoted in Furness 2007:310). In the United States, some 20 pedestrian advocacy groups have formed a coalition called America Walks, and citizens consisting of walkers and bicyclists has pressured cities to create greenways and bike paths tin communities around the country (Kay 1997).

Even though cycling is generally considered a healthy activity, it is also one that is highly stigmatized particularly by motorists but also pedestrians (Aldred 2013:261). Stereotypes of cyclists include the following points:

- Cyclists often disregard the law by failing to stop for traffic lights or stop signs or ride on narrow sidewalks
- Cyclists may be ignorant of the rules of the road, including failing to use front and rear lights at night or in rainy weather
- Cyclists generally do not undergo training and generally are unlicensed and uninsured

Countries and localities vary on whether cyclists should wear helmets. In Australia, the law dictates that cyclists must wear helmets, even on cycle paths.

Like Critical Mass in North America, the environmental movement in Western Europe has mobilized as a counterhegemonic movement opposed to the automobilization of society by emphasizing the need for people to rely on other forms of transportation, including cycling. Environmentalists in Germany, for example, attempt to promote bicycle riding through city streets to slow or halt traffic. Grass-roots groups in Copenhagen, Amsterdam, and other

Dutch cities have done much to create bikeways, marked paths for cyclists on roads and a "culture of respect" for cyclists (Newman and Kenworthy 1999:206). Copenhagen and the community of Frederiksberg situated within the city limits has 307 kilometres of bike paths and in 2000 "34% of home to place-of-work trips were made by cycle" (Batterbury 2002:203).

However, although cycling constitutes an environmentally friendly mode of transportation as well as a form of aerobic exercise, it will remain a highly dangerous activity as long as streets and highways are filled with fast-moving motor vehicles (increasingly occupied by distracted drivers busily cutting business deals or socializing on car telephones as well as mobile phones and thus endangering lives even further) and exhaust fumes. Gotschi, Garrard and Giles-Corti 2016:60) argue that "[c]ycling can result in greater exposure to air pollution because, first, air pollution concentrations are higher in traffic than most other places we spend time . . . , and second, increased ventilation rates lead to higher volumes of inhaled air." Parked cars can pose a danger to cyclists, particularly when drivers suddenly open their doors unaware of or oblivious to a passing cyclist.

Finally, at a possibly more mainstream level, various public transportation associations and groups have come into existence at the international, national, and local levels. The International Association of Public Transport (2014:1) is an "international network for public transport authorities and operators, policy decision-makers, scientific institutes and the public transport supply and service industry" consisting of some 1,300-member organizations representing 92 countries. In its declaration on climate leadership, it states: "Ambitious and visionary actions and strategies are essential to change radical current mobility patterns and avoid dangerous climate change. Cities and governments have a crucial role to play in this and public transport needs to be put forward to tackle the urban mobility challenges currently faced by our cities instead of continuing with the construction of new highways and encouraging car use" (International Association of Public Transport 2014:1). UTIP aims to double the market share of public transportation worldwide by 2025.

In the early 1990s Germany developed the Avoid, Shift, Improve (ASI) approach to climate change mitigation and the environmental impacts of transportation, including off course automobiles. The Avoid strategy would entail "avoiding the need to travel, e.g. by improved urban planning, TDM or road pricing and e-communication options (mobile phone use, teleworking)," the Shift strategy would entail "shifting travel to the most efficient or clean, e.g. non-motorized or public transport or, for freight, rail or waterborne transport" and the Improve strategy would entail "improving the environmental performance of transport through technological improvements to make vehicles more energy efficient and fuels less carbon intensive" (Baker, Zuidgeest, de Coninick, and Huizenga 2014:339).

CONCLUSION

This chapter examines critiques of automobiles, reformist and non-reformist reforms in addressing the increase of private motor vehicles, the emergence of electric and self-driving automobiles, opposition to cars and highways, and the promotion of public transportation. In contrast to advocates of electric and driver-less automobiles, Furness (2010:208) boldly declares that at least in cities automobiles of any sort have no long-term future, with the larger question being "when automobility will be obsolete. Completely transcending automobiles would be virtually impossible in the best of all worlds, but greatly reducing their numbers might be possible, although probably not under the parameters of a capitalist world system which relies upon them for profit-making and economic growth. The more profound issue is how to overcome a heavy reliance or dependence upon them, one which I explore in the next chapter. As Freund and Martin (1993:5) observe, "[o]ur task is not to eliminate but to reduce our auto-dependence and to move away from auto-centered transport systems toward systems that feature a greater variety of modalities."

It is not enough to make automobiles and other motor vehicles more energy efficient and environmentally friendly. In the case of automobiles, it is imperative to seek ways to transcend them as much as possible, but such a shift ultimately will dictate transcending the capitalist world system with its commitment to profit-making and continual economic growth which drive a treadmill of production and consumption that places more and more automobiles on a fragile planetary ecosystem. The creation of alternative forms of mobility and transportation would have to be part and parcel of implementing such a transformation. Bongardt et al. (2013:41–43) envision two possible scenarios for the world's passenger transportation system:

Scenario 1—A passenger system where the level of motorization characteristic of North America or Europe becomes the norm in developing countries by 2050

- The global fleet of automobiles increases to 2–3 billion, with a growing proportion of these cars being electric or fueled by alternative fuels
- GHG emissions continue to rise and the global community has come to accept a temperature rise of 4–5°C
- Many cities around the world resemble North American cities, with most people working in CBDs and living in outer suburbs

- High fuel costs limit social interaction and work opportunities
- Many suburbs have become dilapidated and places of social unrest
- Obesity has become a mainstream disease for both developed and developing countries
- Affluent people access the amenities of suburban life by automobile
- Motorists spend much of their traveling time in heavy traffic
- Automobiles and airplanes account for most intercity and non-urban transportation
- Trains and coach services are unable to offer competitive service due to largely dispersed populations
- Cities are dominated by roads and parking areas of various types, contributing to a profound heat island effect
- People living in hot summer conditions due to global warming find living and working in cities extremely difficult

Scenario 2—A more sustainable system in which reliance upon automobiles would follow the current norm in Europe and Japan

- The global fleet of automobiles is limited to about 1 billion, with the majority of these being small and lightweight and powered by electric, plug-in hybrid or alternative fuel engines
- Greenhouse gas emissions have been reduced to less than 20 percent of 1990 levels
- Global warming has been kept in check to around 2°C, thus avoiding the worst forms of erratic climatic events
- Intercity trains and long-distance buses account for a large portion of interurban transportation
- Regulations have halted the increase of air travel
- Rural areas have been revitalized through sustainable agricultural practices, a mix of telecommunication, low-carbon vehicles, and other sustainable transportation modes
- Conventional bicycles and electric bicycles account for much of urban transportation and contribute to a healthy population
- Cities have become more compact and outer suburban shopping malls and other facilities are limited
- Green spaces and tree-lined streets absorb rain and greatly reduce the greenhouse effect
- Buildings and streets are designed to allow air flow and mitigate air pollution

John Urry envisions four scenarios by which to sidestep through the cracks in the present automobile system. The first of these scenarios would entail development of the "fast-mobility" city which would permit people to move within cities and between cities very rapidly, such is supposedly becoming reality in Shanghai, Dubai, Qatar, Hong Kong, Rio de Janeiro, Seoul, and Singapore, all of which are situated in developing countries. Fast mobility cities would also be "increasingly orbited by many kinds of manned and unmanned aerial vehicles" (Urry 2016:139). Rather than relying on fossil fuels, a transport vehicle of a different sort in the fast-mobility city would be hydrogen-powered. However, even if hydrogen-based vehicles would be technologically feasible, the vehicles operating and the infrastructure for them to do so in the fast-mobility city would require massive amounts of resources, many of which exist in limited supply and would damage the environment in the sheer process of extracting them.

The second scenario is the Digital or smart city in which "there is widespread substitution of physical movement of objects and people by many modes of digital communication and experience" (Urry 2016:142). People would develop a new consciousness in which they do not feel that they must interact with people as much face-to-face but accept digital communication as an alternative mode of interaction. Digital lives would greatly reduce the need for traveling to other places, whether in one's city or in far-away sites. Furthermore, people "would say that they had visited a particular place even though they only 'travelled' there digitally" (Urry 2016:143).

The "liveable city" as the third scenario would entail replacing high-energy mobility motor vehicles and suburban sprawl with a "smaller scale system of neighbourhoods" reliant on bicycles and personal vehicles which "would be electronically integrated through information payment systems and physical access and connecting with collective forms of transport" (Urry 2016:147). Car-sharing schemes would replace individual or family ownership of automobiles. Many cities will be partly or wholly free of automobiles. The liveable city would entail higher-density living and work sites and educational institutions situated closer to home. Furthermore, there would be a "huge scaling of number of 'international' students around the world" and the provision of products and their repair closer to home, resulting in a "systemic reduction in distances travelled by people, objects, goods, and money" (Urry 2016:148). Social interaction would be greater at the neighbor level than it generally is the case in large, medium-sized and even small cities.

The fourth scenario is the Fortress City in which developed societies "break away from the poorer into fortified enclaves" with their more affluent sectors living in gated and armed encampments. In contrast, the poor would live in "wild zones" (Urry 2016:149). Oil, gas, and war shortages along with periodic wars would negatively impact upon energy production, mobility, and communication. Automobiles, trucks, bicycles, computers, and phone systems would be recycled and repaired at the local level. In this dystopian world, life in the Fortress City "would be conducted with the continuous spectre of warfare, the militarization of young men and the raping of women and girls" (Urry 2016:115).

In his assessment of his four scenarios, Urry argues that the Fast City Future is "relatively unlikely," in part because technologically hydrogen fuel-cell technology may be an inadequate means of powering vehicles. The Digital City may develop in certain affluent places where digital corporations would dictate social governance. The Liveable City is most likely to develop in developed societies impacted by catastrophic climate change and/or a global recession where a powerful social movement through protest, persuasion, and example paves the way for its actualization. In the event that none of these future scenarios materializes, the Fortress City is the most likely future scenario, especially given that "[s]ignificant parts of the world are already, of course, 'fortress cities,' and elements of this system are well established" (Urry 2016:155).

Obviously, the hegemony of automobiles has started to crack. As I have indicated in this chapter, proposals to transcend automobiles or at least to mitigate their worst impacts are occurring within the bowels of global capitalism itself. Most of the solutions that I have mentioned in this chapter fit into the rubric of Gorz's reformist reforms but some of them have the potential of constituting non-reformist reforms or system-challenging reforms. In the next chapter, I touch further upon non-reformist reforms that have the potential of not only transcending automobility but also global capitalism itself with a system based upon social justice, democratic processes, environmental sustainability, and a safe climate. Part and parcel of this transitional process would be the creation of sustainable public transportation.

Chapter Eight

Creating a Sustainable Transportation System within the Context of an Alternative World System

Despite its many environmental, social, and health deficits of motor vehicles and growing opposition to their presence in modern life, the culture of automobility is alive and well and increasingly spreading to developing countries, thus contributing to the creation of a dystopian world, characterized by gross socioeconomic inequality, authoritarian practices, and a global ecological crisis, particularly manifested in climate change. These developments have prompted Ladd (2008:179) to observe that the "grave crisis of civilization that some pessimists predict (whether precipitated by fuel shortages, global warming, or the collapse of urban order) may, in fact, if it arrives, be the only way to get people out of their cars—if it actually does so." In the long run, the contradictions associated with automobility, including those associated with the environment, settlement patterns, and health, can only be adequately addressed through the creation of an eco-socialist world system, a system based upon meeting human social needs and social justice, democratic processes, a sustainable environment, and a safe climate, a topic which I have discussed in some detail elsewhere.

In a Next System Project interview on March 24, 2016, renowned linguist and social critic Noam Chomsky argues that during the Global Financial Crisis when the Obama administration bailed out the automobile industry, there was another option that could have been pursued:

> The other choice was to hand the system over to the workforce, have it democratically controlled and managed, and have the production oriented toward what the community needs. We don't need more cars. We need effective mass transportation for lots of reasons. You can take high speed trains from Beijing to Kazakhstan, but not from Boston to New York. Infrastructure is collapsing: it has a horrible effect on the environment. It means spending half your life

in traffic jams. This is implicit in market systems. A market gives you choice among consumer goods, say a Ford and a Toyota. It doesn't give you a choice between an automobile and a decent mass transportation system. (Next System Project 2016)

Humanity is confronted by two imperatives: (1) how do we live in harmony with each other on a fragile planet of limited resources which have become unevenly distributed and (2) how do we live in harmony with nature, particularly as humanity lurches forward into an era of potentially catastrophic anthropogenic climate change which to a large degree is a by-product of the capitalist world system. Social systems, whether they exist at the local, regional, or global levels, do not last forever. Capitalism as a globalizing political economic system committed to profit-making and continual economic growth has resulted in a treadmill of production and consumption which is heavily dependent upon fossil fuels, resulting in greenhouse gas emissions that in turn drive climate change. While capitalism has produced numerous impressive technological innovations, some beneficial and others destructive, which are very unevenly distributed, it is a system fraught with numerous contradictions, including growing social disparities within most nation-states; authoritarian and militarist practices; depletion of natural resources; environmental degradation, including global warming and associated climatic changes; species extinction; and population growth as a by-product of poverty. Even more so than in earlier stages of capitalism, transnational corporations and their associated bodies, such as the World Bank, the International Monetary Fund, and the World Trade Organization, make or break governments and politicians around the world, although the extent to which this is true varies from country to country. Although capitalism has been around for about 500 years, it manifests so many contradictions that it has become increasingly clear that it must be replaced by a "next system" or an alternative world system—one oriented to social parity and justice, democratic processes, and environmental sustainability, which includes a safe climate.

ECO-SOCIALISM

It is imperative that progressive people reinvent the notion of socialism by recognizing that we live on a planet with limited resources which must be more evenly, indeed much more evenly, distributed to provide everyone with enough, but not too much. Eco-socialism is a perspective that has emerged over the course of the past 35–40 years which endeavors to envision an alternative world system which would manifest itself in different ways in different places around the globe (Foster 2009; Loewy 2015).

Eco-socialism entails the following principles:

- An economy oriented to meeting basic social needs—namely adequate food, clothing, shelter, education, health, and dignified work
- A high degree of social equality
- Public ownership of the means of production
- Representative and participatory democracy
- Environmental sustainability

Eco-socialism rejects a statist, growth-oriented, productivist ethic and recognizes that humans live on an ecologically fragile planet with limited resources that must be sustained and renewed as much as possible for future generations.

The vision of eco-socialism closely resembles what world systems theorists Terry Boswell and Christopher Chase-Dunn (2000) term *global democracy*, a concept that entails the following components: (1) an increasing movement toward public ownership of productive forces at local, regional, national, and international levels; (2) the development of an economy oriented toward meeting social needs, such as basic food, clothing, shelter, and health care, and environmental sustainability rather than profit-making; (3) the eradication of health and social disparities and the redistribution of human resources between developed and developing societies and within societies in general; (4) the curtailment of population growth that in large part would follow from the previously mentioned conditions; (5) the conservation of finite resources and the development of renewable energy resources; (6) the redesign of settlement and transport systems to reduce energy demands and greenhouse gas emissions; and (7) the reduction of wastes through recycling and transcending the reigning culture of consumption. Eco-socialism constitutes what sociologist Erik Olin Wright (2010) terms a *real utopia*, a utopian vision which is achievable but only through much theorizing and social experimentation. As the existing capitalist world system continues to self-destruct due to its socially unjust and environmentally unsustainable practices, democratic eco-socialism seeks to provide a vision to mobilize human beings around the world, albeit in different ways, to prevent on-going human socioeconomic and environmental destruction.

While Stalin adhered to the notion of building "socialism in one country," what developed in the USSR for complicated reasons, historical and social structural, internal and external was the creation of a highly authoritarian and draconian social system that made a mockery of the notion of Marxian socialism. In keeping with Trotsky's notion of the "permanent revolution," the creation of socialism requires a global process, the beginnings that we may be seeing rekindled in the guise of the Bolivarian Revolution in Latin America (albeit an experiment with numerous contradictions) and the emergence of new left parties in Europe, particularly Syriza in Greece which came to power earlier in 2015 and the *Die Linke*, the farthest left party in the German *Bundestag*. As global capitalism continues to find itself in economic and ecological crises as it lurches into the twenty-first century, humanity finds itself with the challenge of how to shift gears from an on-going trajectory of human and planetary destruction. As the existing capitalist world system continues to self-destruct due to its socially unjust and environmentally unsustainable practices, democratic eco-socialism provides a radical vision to mobilize people around the world to struggle for the next system.

Anti-systemic movements are sure to be a permanent feature of the world's political landscape so long as capitalism remains a hegemonic political-economic system. Various anti-systemic movements, particularly the labor, ethnic and indigenous rights, women's, anti-corporate globalization, peace, and environmental and climate movements, have an important role to play in creating a socio-ecological revolution committed to both social justice and environmental sustainability. Anti-systemic movements are a crucial component of moving humanity to an alternative world system, but the process is a tedious and convoluted one with no guarantees, especially considering the disparate nature of these movements.

Transitional Systemic-Challenging Reforms That Potentially Can Move the World beyond Motor Vehicles

Historically Marxists or socialists have engaged in intense debates as to whether the transition from capitalism to socialism will occur vis-à-vis revolutionary change or more gradual change in the form of reforms in various parts of the world. Revolutions involve more sudden social and radical social transformations and are often associated with much violence, as was the case with the American, French, Chinese, and Cuban revolutions, but ironically was not the case for the Bolshevik Revolution in Russia in October 1917.

The transition toward an eco-socialist world system is not guaranteed and will require a tedious, even convoluted path, one in which anti-systemic movements will have to play a central role. Marx viewed blueprints as a distraction from the political tasks that needed to be undertaken in the present moment and it is important to note that these indeed are paramount. But history tells us that there always will be immediate struggles which must be addressed. I often find that when people ask me what it would take to make a transition to a eco-socialist world system, they are seeking some basic guidelines on how to move forward beyond merely bumbling along haphazardly a step at a time.

While not seeking to create a blueprint per se for creating an alternative world system which will be manifested in different ways in the many societies around the world, I have proposed the following systemic-challenging reforms to facilitate a transition from the present existing capitalist world system to a democratic eco-socialist world system: (1) the creation of new progressive, anti-capitalist parties designed to capture the state; (2) the implementation of greenhouse gas emissions taxes at the sites of production that include measures to protect low-income people; (3) increasing public ownership, socialization, or nationalization in various ways of the means of production; (4) increasing social equality within nation-states and between nation-states and achieving a sustainable global population; (5) the implementation of workers' democracy; (6) the creation of meaningful work and shortening the work week; (7) achieving a net zero growth economy; (8) the adoption of energy efficiency, renewable energy sources, and green jobs; (9) the expansion of public transportation and massive diminishment of a reliance on private motor vehicles and air travel; (10) the implementation of sustainable food production and forestry; (11) resistance to the culture of consumption and adoption of sustainable and meaningful consumption; (12) the implementation of sustainable trade; and (13) the implementation of sustainable settlement patterns and local communities (Baer 2018). In this book, I focus particularly on transitional reforms that touch upon transportation and more specifically automobility.

My transitional steps constitute loose guidelines for shifting human societies or countries toward eco-socialism and a safe climate. It is important to note that both these phenomena will entail a global effort, including the creation of a progressive global climate governance regime. My litany of proposed transitional reforms is a modest effort to contribute to an on-going dialogue and debate as to how to move forward from the present impasse in which the world finds itself today. The application of my suggested transitional reforms must be adapted for many countries, both developed and developing, around the world. Furthermore, my suggested transitional reforms do not exhaust possible transitional reforms necessary for creating an alternative world system. Because I have discussed these transitional systemic-change reforms in detail elsewhere (Baer 2018), for purposes of this book I will touch upon those which most directly relate to the culture of automobility and the need to create a sustainable transportation system.

Public Ownership of the Means of Production

In an era of increasing privatization of social and health services, even military activities and prisons, raising the specter of public ownership, nationalization, or socialization of the means of ownership is taboo in conventional economic and political circles. Privatization is often justified in terms of economic efficiency. While state or government enterprises or services can be terribly inefficient for complex reasons, this does not necessarily have to be the case. There are numerous examples of publicly owned enterprises that operate relatively efficiently. Public ownership could consist of various social arrangements, including state ownership, worker-owned enterprises, and cooperatives.

Herman Rosenfeld (2009), a member of the Canadian Socialist Project and the Greater Toronto Workers' Assembly, advocates a comprehensive socialist approach to the search for solutions to the economic and environmental crises generated by the North American automobile industry. He argues that the global financial crisis of 2008–2009 pointed out the need to restructure the North American automobile industry which avoided bankruptcy only due to government bailouts. Rosenfeld (2009) maintains that the private welfare state needs to be replaced by a set of strengthened, democratically administered, universal public programs; the banking and finance sector should be nationalized and socialized and operated by democratic bodies; the regulation of auto production and trade; and that much of the productive capacity presently used to manufacture cars should be redirected to produce other goods and services, including those presumably related to public transportation. Rosenfeld (2009) also notes that efforts to address climate change and the environmental crisis in general would require a shift to smaller vehicles reliant on non-fossil fuels; a strong emphasis on public transportation; the development of alternative sources of fuel and energy; and the creation of new ways of working and living. He believes that communities must be restructured to defend the right of workers to decent jobs and direct control over the production process. Rosenfeld proposes a socialist approach to resolving the automobile crisis which entails the following components:

- Replacing the existing "private welfare state" with a "set of strengthened, democratically, universal public programs"
- Nationalizing and socializing the banking and finance sectors and having democratic bodies administer them
- Regulating automobile production and trade
- Shifting to a society with fewer private and commercial motor vehicles in order to address climate change and the ecological crisis which includes the following steps
  - Production of newer, smaller vehicles that rely on non-fossil fuels
  - Expansion of mass transportation
  - Development of alternative sources of fuel and energy
  - Development of "new forms of living, working, and enjoying recreation time"
- Shifting the production of automobiles to other goods or services

Lars Henriksson (2013:82) elaborates upon the last point above and advocates a "third road" for the declining Swedish automobile industry beyond the two conventional approaches of either allowing the industry to die, as has been the case for the Swedish shipbuilding and textile industries, or the state propping it up with subsidies, proposing that the Swedish state nationalize the "industry and convert it to create safe jobs and a production that can help us move away from the fossil fuel economy."

While it would necessarily shift developed societies away from automobiles, Konczal (2014) argues "car sharing" companies, particularly Uber, that exploit their drivers by requiring them to pay for their automobiles and maintenance and fuel need to be socialized by the drivers themselves. In the United States, some 160,000 people work for Uber, with only about only 4,000 being regular employees (Streeck 2016:26). As a worker-owned cooperative, a car sharing company would see all profits going to the drivers who decide how to run their enterprises. Konczal (2014) states: "Uber and the rest of the "sharing economy" companies will try to close the door behind them, either by putting their workers in binding contracts or by lobbying government officials to build their own set of industry protections. But a transition to workers' owning their firms is necessary, economically smart, and one way for workers to gain power in the digital age."

Energy Efficiency, Renewable Energy Sources, and Green Jobs

A crucial question is how much energy, regardless of the source, does humanity need. Given the demands of global capitalism to continually expand, under a business-as-usual scenario, humanity will need more and more energy to feed its treadmill of production and consumption and population growth. In a steady-state economy, energy requirements could theoretically level out or even eventually decline. Presently, the average automobile wastes over three quarters of the fuel it utilizes (Williams 2010:199). The US federal government implemented Corporate Average Fuel Economy (CAFÉ) standards for automobiles under the Energy Policy and Conservation Action in 1975. The Obama Administration issued joint rules on vehicle fuel economy and greenhouse gas emissions for model year 2012–2016 in the case of passenger cars and light trucks, model year 2014–2018 in the case of medium- and heavy-duty tricks, and model year 2017–2025 again in the case of passenger cars and trucks, but the fate of these limited regulations remains in the balance given the fact that the current president of the United States, namely Donald Trump, is a staunch climate denialist (Yacobucci, Canis, and Lattanzio 2012:1).

There has been much discussion about electric motor vehicles, including private electric automobiles, as being more energy efficient, producing less air pollution and greenhouse emissions than motor vehicles powered by the internal combustion engine. This would be true if the electricity could be drawn from batteries or large solar power plants but not necessarily if the electricity would be drawn from coal-fired power plants. There is much room that could be made for making automobiles more energy efficient, a topic which has received much consideration but lies beyond the purview of this book. Electric buses, trains, and trucks for purposes of at least local deliveries would be part of the mix of a sustainable transportation system but merely substituting electric automobiles for conventional ones would not address a growing congestion problem, particularly in developing societies.

Although there has been a push for greater fuel efficiency in the United States and other developed countries, there has also been a trend toward larger and larger passenger motor vehicles which emit more greenhouse gas emissions. Ultimately, technological and energy efficiency innovations, such as electric cars, hybrid cars, smart cars, and automated cars (if they should ever come into existence), are not long-term solutions to the ecological crisis unless they are coupled with a push for a simpler way, particularly in developed societies but also among the affluent sectors in developing societies.

Energy efficiency is often hailed as a mechanism for transition to a green-energy economy, but due to the Jevons Paradox or the "rebound effect," increased efficiency in capitalist countries is associated with increased economic growth and consumption, thus in essence canceling out the benefits of energy savings. This is not to say that energy efficiency is not a desirable goal but to ensure environmental sustainability, it has to be coupled with a steady-state or zero-growth global economy, which would be part and parcel of a democratic eco-socialist world system.

A shift to renewable energy sources, particularly solar, wind, geothermal energy, and possibly ocean wave energy, constitutes a significant component of climate change mitigation. A planned centralized economy has the potential to facilitate the transition to renewable energy sources, which could be used in transportation modalities. Aside of the matter of renewable energy sources, eco-socialism needs to grapple with developing a "socialist technology." The component parts of a socialist technology to some extent already exist in capitalist societies but are not actively promoted by capitalism because they are not as profitable. The technology already exists to make products that endure for a long time rather than are manufactured in such a way that they will break down rapidly, a case of built-in obsolescence. Bicycles, smaller cars, and trains, trams, and buses as opposed to large cars, all could be part of a socialist or an appropriate technology. Finally, a shift to eco-socialism will entail creating "green jobs," ones that are not only environmentally sustainable but also cater to people's social, educational, recreational, and healthcare needs. The creation of green jobs must be accompanied by a "just transition," which means retraining displaced workers, including ones from the private motor vehicle industry for jobs in nationalized or socialized enterprises that manufacture trains, trams, and buses that could be part and parcel of an integrated public transportation system.

Sustainable Public Transportation and Travel

The negative environmental impacts of private motor vehicles will require a drastic shift to sustainable public transportation. Like the term "sustainable development," sustainable mobility and transportation or transport has come to mean different things to different parties. Even automobile companies and the OECD, bodies that are fully committed to a capitalist and neoliberal agenda, have embraced these notions. Nicholas Low, a Professor of Urban and Environmental Planning at the University of Melbourne, proposes two principles of "sustainable transport" that strike me as congruent with a democratic eco-socialist perspective on both social justice and environmental sustainability, albeit with some qualifications. His first principle of sustainable transport states: *"For an urban transport system to be sustainable it must be one in which the carbon emissions from its operation, and embodied in new infrastructure construction, are reduced to a level compatible with a global temperature rise from pre-industrial levels of no more than 2 degrees Celsius"* (Low 2013:70–71). In recent years, however, some climate scientists, such as James Hansen, have come to make a case that a 1.5° C limit above pre-industrial levels is needed to achieve a safe climate. Low's (2013:70) second principle of sustainable transport states: *"For a sustainable urban transport system also to be fair it must be one in which per capita carbon emissions are approximately the same in all cities worldwide."* His second principle calls for social justice and parity in what is a highly socially stratified capitalist world system. From my perspective, achieving this would require a radical transformation that would include a high degree of social parity and redistribution of income, wealth, and resources globally.

As Paterson (2014:571) observes, "addressing climate change adequately involves radical reductions in certain forms of mobility," particularly automobiles and airplanes but also in the transportation of food and goods. Michael Loewy (2015:34) argues that the transition to eco-socialism would entail a massive expansion of public transportation and the drastic reduction in reliance on private automobiles. Furthermore, "in transition to a new society, to drastically reduce the transportation of good by trucks—responsible for terrible accidents and high levels of pollution—replacing them with rail transport or what the French call *ferroutage* (trucks transported in trains from one town to another" (Loewy 2015:35).

A new urbanism that seeks to make cities more liveable and environmentally sustainable has emerged around the world and has begun to permeate urban planning. Various cities around the world, including Singapore, Hong Kong, Zurich, Copenhagen, Freiburg (Germany), Vancouver, Toronto, and Boston, are encouraging residents to rely more on public transportation, including trains, trams, and buses. A global movement to make inner cities car-free has emerged in recent years. Car-free development requires enormous political will which generally does not emanate from corporate interests, mainstream politicians, or even urban planners. It is more likely to emanate from progressive political parties, environmental NGOs, and anti-automobile activists.

Sustainable transportation would entail many other measures, such as limiting the use of cars as much as possible, making them smaller and more energy efficient, and even banning four-wheel drives or sports utility vehicles (SUVs), except in special circumstances (such as in the outback and rugged mountainous areas) and drastically limiting air travel.

Newman, Kenworthy, and Glazebrook (2013:272) delineate twelve advantages of rail travel in urban areas:

- "Lower per-capita traffic congestion costs"
- "Lower per capita private passenger transportation energy use"
- "Lower per capita emissions from the transportation-sector"
- "Lower per capita traffic fatalities"
- "Lower per capita consumer transportation expenditures"
- "Higher per capita transit service provision"
- "Higher per capita transit ridership"
- "Higher transit commute mode split"
- "Lower transit operating costs per passenger mile"
- "Higher transit service operating cost recovery"
- "Lower CBD parking per 1000 jobs"
- "Better overall urban design in the city especially through Light Rail Transit (LRT) systems"

As of 2007, LRT systems existed in numerous European cities, including five in Belgium, 56 in Germany, eleven in France, seven in the UK, seven in Italy, 14 in Poland, 14 in Romania, and seven in the Czech Republic (Newman. Kenworthy, and Glazebrook 2013:274). Cities in several Middle Eastern countries and Asian countries have plans to develop metro systems. Unfortunately, public transportation can be quite expensive. Brisbane in Australia reportedly has some of the most expensive public transportation fares in the world, which raises the issue of transportation justice as a sub-type of social justice or social parity (Li, Dodson, and Sipe 2015). Should public transportation be fully government subsidized or should fares be means-tested until such time that a more socially equitable social system exists? Would free fares encourage people in CBDs to overuse public transportation and thus not walk short distances more? There are no easy answers to such questions and to some degree they fall outside the scope of this book. Nevertheless, the larger issue is shifting people away from private motor vehicles as much as possible and toward alternative forms of mobility, be it public transportation, cycling, or walking.

Much thought is being given to the best form of public transportation, such as train, tram, or bus, in urban areas, depending on the situation. Furthermore, there is the issue of connecting small towns and rural areas with cities. Measures will need to be taken to connect rural communities to urban communities and to provide public transportation, perhaps in the form of regularly scheduled mini-buses in rural areas. Furthermore, it would be possible to reinstate passenger rail service that serviced rural communities in both North America and Australia at a time in the past when their respective populations were considerably less than today. While shifting from cars to public transit, particularly intercity-trains and suburban trains and trams or light-rail systems, would serve to diminish greenhouse gas emissions, these modes of transportation are not panaceas. While there is much discussion about high-speed passenger rail substituting for cars or plane travel, according to Todorovich and Burgess 2013:145), various studies indicate that the "direct benefits from high-speed rail in terms of overall energy and emissions may be modest." Additionally, "these analyses also neglect the indirect impacts in terms of land-use and city-entering, which may be large and are difficult to measure and attribute. It is our view that these indirect benefits may be more important than any direct reduction in energy utilization as passengers choose high-speed rail over alternative travel modes" (Todorovich and Burgess 2013:145).

Public transportation in the highly automobile-dependent United States bottomed out in the early 1970s, with Americans slowly turning to it since then (Foster 2003:194). Peter Newman and Christy Newman (2012:360) report: "Recent data from US and Australian cities show that car use per person has been going down since 2004, and that public transport has been dramatically increasing. The first assessments of this suggest that something structural is happening, that younger people in particular are coming back into cities rather than choosing car dependence. Indeed car ownership among teenagers in the United States has dropped from fifteen million to ten million. Is it possible that a change in car culture is underway?"

Ironically, there was a time when cities, particularly North American and Australian cities, had better public transportation systems than they do now, which included extensive trolley or tram systems. Various large Australian cities, including Sydney, Melbourne, Brisbane, Perth, and Adelaide have extensive urban train systems that could be improved upon if funds were redirected from highway construction to public transportation. Unfortunately, as Dodson and Sipe (2008:79) observe, "Australia's rail networks suffer from extraordinarily poor management, and there is negligible integration of bus systems with the rail links." Fortunately, various US and Canadian cities have over the past two decades or so developed "light rail systems." These include Baltimore, Buffalo, Cleveland, Pittsburgh, Minneapolis, St. Louis, Dallas-Fort Worth, Denver, Salt Lake City, San Diego, Los Angeles, San Jose, San Francisco, Sacramento, Portland (Oregon), Seattle-Tacoma, New Orleans, Newark, Toronto, Edmonton, and Calgary (Black 2010:72). Portland in particular has become an exemplar on the benefits of shifting away from automobiles toward greater reliance on a light rail and buses, resulting in a doubling of transit ridership in the "past 20 years (totalling 100 million rides in 2013)" (Holland and Wei 2016:293). Unfortunately, automobiles continue to reign supreme in Portland as they do even more so in most other US cities.

While many European cities eliminated their tram systems in the 1950s and 1960s, a major revival of light rail has occurred in many European cities over the past three decades or so: "No fewer than 65 cities built new or expanded light-rail systems between 1980 and 2007, bringing the total number of European cities with light rail at that time to over 160. Further growth in light rail has continued since then, particularly in France, Spain, and Portugal" (Newman and Kenworthy 2015:9–11). New and expanded suburban metro systems have been added or are being constructed in many large European cities, including Paris, Madrid, London, Vienna, Stockholm, Munich, and Frankfurt.

While public transportation in most cities in developing countries consists primarily of bus services, "there are a small number of big cities with suburban train services (for example Mumbai and Chennai in India) and some with metro rail systems (for example Beijing, Shanghai and Guangzhou in China; and Delhi and Kolkata in India)" (Pendakur 2011:208). In addition to the metro systems in Shanghai and Beijing, between 1997 and 2013, China opened metro systems in 17 additional cities and metros are under construction in another 20 additional cities (Newman and Kentworthy 2015:12).

For a variety of reasons, including poor scheduling or limited routes, buses are frequently underutilized. For example, as a frequent public transit user in Melbourne, while I have found that trains and trams are well used, often requiring some passengers traveling at peak hours to stand, buses which often serve to connect train and tram lines, often are greatly underutilized. According to Paul Mees (2010:38), "A bus with half a dozen passengers will be no more efficient, in greenhouse terms, than if the passengers travelled in cars at average occupancies." Conversely, various Latin American cities, such as Curitiba in Brazil, have created rapid, low-cost high-frequency bus systems that utilize bus-only lanes and feeder buses that link with the larger bus network (Metz 2010).

One of the issues that has received little consideration is the danger that women, particularly older women, feel on public transportation. Based upon a survey conducted among women in the Northeast of England, Dobbs (2005:274) reports: "[P]ublic transport does not appear to provide women with the security they need. On the contrary, women talk extensively about the problems of using public transport itself, suggesting that they are afraid of travelling on public transport and that they find stops and stations 'threatening and intimidating.' They also refer to the way in which public transport leaves them isolated within an increasingly hostile environment, essentially 'making it impossible for [them] to travel safely."

I have frequently heard women over age 50 or so in Melbourne voice similar sentiments, particularly with regards to riding the Metro, the extensive suburban train system, that serves metropolitan Melbourne. While the topic lies beyond the purview of this book, the inequities characteristic of capitalist societies creates massive alienation and poor economic opportunities that prompts particularly young males to turn to petty theft and even violent sexual acts. In an eco-socialist world, such gross socioeconomic inequality would be eradicated and hopefully moving about in public places, including accessing public transportation stations and stops, much safer.

In capitalist societies, "time is money," and this dictates rapid movement between places. Indeed, as Marx observed, capitalism engages in time-compression. Freund and Martin (1993:172) state: "Accelerated movement and speeded-up time are archetypical features of both industrial and post-industrial landscapes. Speeded-up time schedules and widely dispersed sites of human activity are some of the objective correlates of experiences such as placelessness, community deterioration, and ecological degradation. Autos and trucks provide greater flexibility than rail alone, but they do so at greater social and environmental costs." Conversely, in a more leisurely paced world based on eco-socialist principles, people might find slower train travel—although faster than presently exists in most parts of North America and Australia—to be a time to slow down by reading, chatting with fellow passengers, enjoying the passing countryside, reflecting, and even sleeping. A more sustainable form of vacationing or holidaying would entail trips much closer to home, by train or bus, if possible rather than to distant places either by plane or car.

To compete with air travel, there has been much discussion about the need for high-speed trains, although these would not be a panacea, particularly if the train derives its energy from diesel fuel or electricity generated by coal-fired powered power plants, given that the faster the train goes, the more energy it consumes, and the more CO_2 emits per distance to overcome the greater wind resistance. Reportedly a "train traveling at say, 400 kph, requires 4 times as much energy as one traveling at 200 kph and 16 times as much as one traveling at 100 kph" (Hillman 2012:23)

The limits to resources and issues of environmental sustainability require re-examination of a growing expectation of unrestricted mobility. The limits to resources and issues of environmental sustainability require re-examination of a growing expectation for unrestricted mobility. There is a need to emphasize local mobility over rapid mobility covering long distances, particularly by airplane and automobile. Furthermore, more environmentally friendly modes of transportation for long distance travel would be by train or bus. In an era of ecological and climatic crises, humanity needs to transcend a capitalist-driven emphasis on seeing beautiful places in faraway lands and focus on the natural beauty closer to home.

During my 15 years of residence in Australia, I have found it interesting how many Australians I have met who have traveled to distant destinations in Europe, North America, and Asia but have not visited places closer to home, even within their own state, let alone other parts of Australia or New Zealand. While train travel in Australia is much more sporadic than it is in either European countries or Japan, Victoria, New South Wales, and Queensland have relatively extensive passenger railway networks, which at least in the cases of Victoria and New South Wales, connect with coach services that provide connections to many country towns within their states. Indeed, NSW Train runs regular daily trains between Sydney and Brisbane, Sydney and Melbourne, and Sydney and Canberra. Western Australia has a state railway that connects Perth to Bunbury to the South and Perth to Kalgoorlie. The Great Southern Railway is a private consortium that operates three passenger lines: (1) the Indian Pacific (Sydney–Adelaide–Perth), (2) the Overland (Melbourne–Adelaide), and (3) the Ghan (Adelaide–Alice Springs–Darwin). Unfortunately, these trains, particularly the Indian Pacific and the Ghan, are largely expensive tourist operations, and have rather sporadic timetables. Socialization or nationalization of the Great Southern Railway could pave the way for more frequent runs and lower fares.

Cheap package holidays by airplane could become thing of past. A simpler way would also entail a disposal of or minimizing the use of private motor vehicles and reliance on alternative modes of transportation, including simple walking and cycling. Airships would constitute a form of slow travel given that they travel at speeds of 150 to 200 kilometers per hour. Transoceanic ships could make considerable use of wind power through the use of kites or solid sails. While the world is not in need of empires of any sort, bear in mind that Britain and Spain, as Heinberg (2015: 137) observed, "managed to build trans-oceanic empires using sails." Teleconferencing also has the potential to eliminate or reduce much air travel for purposes of conducting business or attending conferences.

SUSTAINABLE SETTLEMENT PATTERNS AND LOCAL COMMUNITIES

Modern cities have evolved following, in large part, the dictates of capital with its need for manufacturing, financial, commercial, distribution, and communication centers, as well as the administrative demands of government bureaucracies. As cities have grown, they have gobbled up precious farmland and natural areas. Overall cities are energy-intensive places on various counts, including in the operation of office buildings, industries, residences, shopping centers, recreational facilities, restaurants, educational institutions, hospitals, transportation systems, highways, parking lots, airports, etc.

While advocates of green cities often argue that urban density can serve to foster environmental sustainability, the ecological and carbon footprints of cities vary considerably between cities in developed and developing countries but also within cities, depending upon their residential patterns (e.g., McMansions versus slum dwellings) and modes of transportation (e.g., a city with an excellent public transportation system versus a highly car-dependent one). The ecological and carbon footprints of cities extend far beyond their boundaries because they rely upon resources from a larger hinterland which literally encompasses much of the world. Various proponents of "sustainable cities" who maintain that increasing urban density contributes to environmental sustainability downplay the historical connections between density and economic growth.

Theoretically cities have the potential of becoming much greener than they presently are. During the early twentieth century, various socialists and anarchists pioneered efforts to make cities more liveable both socially and environmentally, such as the Karl Marx-Hof in Red Vienna and the Bauhaus housing experiments in Germany. A new urbanism that seeks to make cities more liveable and environmentally sustainable has emerged around the world. It needs to make a much stronger effort to be socially inclusive and counteract gentrification which marginalizes low-income people. Conversely, in a democratic eco-socialist world, there would be no poor people and differences in income and wealth would not be nearly great as they are in capitalist societies.

The development of green cities constitutes a highly imaginative endeavour, one that will require drawing insights from numerous disciplines and fields, including architecture, building construction, urban planning, transportation planning, and last but not least the social sciences. Based upon a detailed examination of Copenhagen, Petter Naess (2006:265) makes the following recommendations for planning a sustainable and less automobile-dependent urban landscape:

- Most buildings, both residential and non-residential, should be constructed "within existing urban area demarcations"
- Priority is given to building apartment buildings and terrace houses instead of detached single-family houses
- Automobile driving within the city should be restricted, no additional roads or parking facilities will be constructed whereas the existing public transportation system is improved and expanded
- Most of the densification is channeled to areas already affected by technical encroachments, so that the urban green structure can be keep intact as much as possible

Prompted by the recognition that "urban cars and trucks are wrecking global ecosystems and destroying our communities," J. H. Crawford (2002:17) set out to develop a design for carless cities. In this regard, he was inspired by the fact that Venice has always been a carless city, although motor vehicles have emerged on nearby Mestre on the mainland. Unfortunately, gentrification has forced many Venetians to move to Mestre because wealthy people purchased dwelling units in Venice, which many of them leave vacant or have converted into hotels that cater to the tourist trade. Although Venice with its many waterways has few gardens and parks, the absence of automobiles on its streets permits children to play on them and safely walk to school or to visit friends, with older children taking ferries to explore their vibrant city. Crawford (2002:47) reports: "In Venice, virtually all freight is transported by boat, except for a small area near the train station that has direct rail and road service. The narrow waterways and low bridges restrict the size of the freight scows, so their capacity is quite limited. Freight must be transhipped between rail or road and the delivery boats."

Car-free cities constitute a real utopian vision, but one that at least in part exists in places, including Stroget, an extensive section of Copenhagen; most of the center of Freiburg, Germany; and almost all of the village of Zermatt in the Swiss Alps. Most German and Swiss towns have at least a small carefree area in the CBD (Crawford 2002:233). Car-free or near car-free cities could rely upon small, battery-powered electric delivery vehicles or delivery bicycles for stores, offices, and residences (Crawford 2002:195).

There has been much discussion on how to make buildings more environmentally sustainable through the use of green roofs and walls, fritted glazing, solar panels, and more efficient lighting. A green or sustainable city should include medium-density housing, easy access to public transportation, and minimize reliance on motor vehicles. Walkability should be part and parcel of the green city which would allow people to walk as much as possible to their work sites, parks, recreational centers, theaters, shops, and eating places and contribute to a democratized streetscape. Some psychologists have developed the notion of *eco-psychology*, which stresses the need for people, including urban dwellers, to have contact with the natural environment. Eco-villages and cohousing communities, which are increasingly found in urban and rural areas of developed and developing societies, constitute pre-figurative social experiments that potentially are part and parcel of developing more sustainable settlement patterns. Urban eco-villages can reduce car dependence or eliminate it completely if they are close to good public transportation.

Cities should be easily interconnected with other cities visa-à-via trains rather than auto or plane transportation. Also there is the question as to what is the optimum maximum population of a city. Some cities have become so incredibly large that it almost defies the imagination, with the world now having some 25 megacities with populations each of over ten million people: Sao Paulo has 27.6 million, Mexico City 21.6 million, and Mumbai 20.7 million people. Obviously, there is no easy answer to this question because it depends upon the national context and notions of population density.

The new urbanism movement has called for the creation of "new towns" that "would enable their residents to satisfy most of the travel needs—that is, work trips, shopping trips, and recreation—by walking or biking within the town and thus decreasing vehicles on the highway, air pollution, and motor vehicle accidents and injuries" (Black 2010:131). In the United States, examples of "new towns" include Seaside and Celebration in Florida and Kentlands in Gaitherburg, Maryland. Freiburg, a medium-sized city of about 230,000, in many ways has been a long-time implementer of many of the principles of the new urbanism (Banister 2005:122). In the 1950s it adopted a policy of not accommodating automobiles in the same way that many other German cities were doing in the post-World War II boom era and elected to main its historic city center. In 1969 Freiburg opted not only to keep its tram system but to expand it. In 1973 it closed its old town to automobile traffic. Freiburg added 7 kilometers to its tram system in 1985 and developed in 1994 another public transit link for a new residential development of some

12,000 people. In 1989 the city implemented a *Gesamtvekehrskonzeption* (a city-wide transport concept) to reduce reliance on automobiles and increase public transit utilization, which included a park and ride system and bike and ride and walk system. Between 1972 and 2002, Freiburg increased cycling paths from 29 kilometres to 150 kilometers. It restricted motor vehicles to main roads, reduced speed limits, and prohibited non-local traffic from entering residential areas. However, it is important to note that even Freiburg has its automobile-dependent outer suburbs.

A new urbanism that seeks to make cities more liveable and environmentally sustainable has emerged around the world and has come to permeate urban planning. Various cities around the world, including Singapore, Hong Kong, Zurich, Copenhagen, Freiburg (Germany), Vancouver, Toronto, and Boston, are encouraging residents to rely on available forms of public transportation, including trains, trams, and buses. While the United States remains a highly car-dependent country, there are some signs that public transit has started to exhibit a modest upswing. Lane (2013:110) reports: "Conventional bus service has been improved in many cities, and new modes of transit such as express bus and light rail have enhanced this serviceability and image of transit operations. Data from the American Public Transportation Association . . . indicates that mass transit has been increasing in aggregate counts of unlinked passenger trips in the United States since the mid-1990s, while VMT [vehicular motor transportation] seems to have slowed to a halt and, since 2007, actually decreased (Lane 2013:110)." While most US cities constitute a "sea of auto-dependent sprawl" in contrast to most European cities that have efficient and highly utilized public transit systems, it is important to note that the "few American cities, such as New York, San Francisco, Chicago, or Boston, that have good public transportation systems and accompanying density have per-capita fuel/energy consumption and GHG emission that are not only lower than other parts of the United States, but in some cases lower than those in Europe" (Dutka 2013:128). In these US cities, people tend to live in more compact living quarters and rely less on cars for transportation than in most US cities. Conversely, some European cities, such as Madrid, appear to have embraced some of the suburban sprawl characteristic of most large US cities.

While it is obvious that low-density housing, especially when it entails large houses on relatively large plots of land (e.g., the proverbial "quarter of an acre" characteristic of many North American and Australasian cities) has a high ecological footprint, there is a raging debate within urban planning circles as to whether high-density housing and medium-density housing is more environmentally sustainable. With a good public transit system available to residents occupying either type of housing, a city theoretically could overcome or at least mitigate automobile dependence. Mees maintains that automobile dependence in presently existing suburban areas could be greatly reduced. He argues: "In parts of Europe and some other places, the high-quality public transport previously found in dense city centres is being extended to suburbs and even rural areas. Public transport networks which once catered only for peak-hour commuters have been reconfigured to serve cross-city, off-peak and . . . even recreational trips" (Mees 2010:7).

There is also the issue of connecting small towns and rural areas with cities. In Europe, villages, such as in Switzerland, are relatively well-connected to urban areas, but this is not generally the case in North America and Australasia. Thus, given that public transportation is often infrequent or even non-existent, most people have become dependent on cars to connect them with commercial centers, family, friends, and acquaintances. Diana Young (2001) observes that motor vehicles appear to be the most significant Western consumer item for the Anangu people in the South Australian outback in that in their remote community it has multiple purposes, including providing relatives lifts to shop and work, pick up the mail, and pay bills, and constitutes an essential component of fulfilling social obligations. Likewise, the car or the pick-up truck serves a similar purpose for Native Americans living on remote reservations, as any visitor to such communities will observe. Measures will need to be taken to connect rural communities to urban communities and to provide public transportation, perhaps in the form of regularly scheduled mini-buses in rural areas. In terms of countries, Switzerland, admittedly a small country, has the most extensive public transportation system in the world, one that connects urban and rural areas, including steep mountainous terrain. Zurich reintroduced trams "at a time when they were disappearing from many German cities, and new stretches of railway were laid when services were closing down in many other countries" (Welzer 2012:113). Whereas the average person in EU countries makes 14.7 train trips a year, the average Swiss makes 47 train trips a year. Switzerland has managed to develop pulse timetabling strategy that effectively connects cities and small towns with rural and exurban areas. Petersen reports:

> Hourly or half-hourly long distance trains come together (or "pulse") on the hour and half-hour at the network's core stations of Zurich, Basel, and Bern. A similar approach is applied at designated pulse points on local and regional rail networks throughout the country: buses and trains arrive before the pulse time, wait a short time to allow passengers to change between services, and then depart. The pattern repeats every hour (or half hour), and as a result, trains also depart all intermediate stations at repeated times every hour. (Petersen 2016: 178)

Furthermore, it would be possible to reinstate passenger rail service that serviced rural communities in both North America and Australia at a time in the past when their respective populations were considerably less than today.

Cycling

Cycling, along with walking, not only has obvious health benefits but also potentially can contribute to reducing greenhouse gas emissions (Pooley et al. 2012). Ironically, cycling became a commonplace form of mobility throughout various European, African, Latin American, and particularly Asian countries during the first half of the twentieth century and even underwent a minor growth pattern in the United States during World War II, when automobile production was curtailed (Vovanco 2013:32). In recent decades, the bicycle has been dubbed the "vehicle for a small planet." Many cities also are increasing provisions for cycling and walking, including Canberra, the Australian national capital, as forms of active transportation. Various large German cities, including Munich, Hamburg, Berlin, and Frankfurt, have fostered cycling as a cost efficient, appealing, and environmentally sustainable form of public transportation (Lanzendorf and Busch-Geertsema 2014).

Copenhagen has created bicycle right-of-way lanes and has fostered an ethos of respect for cyclists. By 2007 bicycles accounted for more than one third of all transit in Copenhagen (Agyeman 2013:125). Furthermore, much of the Copenhagen city center has been pedestrianized. Xia et al. (2013) maintain the active transport has pronounced health benefits. For instance, public transport usage could increase physical activity per day by range of 8–33 minutes. Regular cycling also has clear-cut health benefits but is an activity which varies considerably between urban areas around the world, ranging from less than 3 percent total travel trips in some cities in the UK, US, and Australia to over 30 percent of trips to work or school in the Netherlands and Denmark and 28 percent in Germany. Schiermonnikog, a national park off the northwest coast of the Netherlands, does not allow visitors to drive their private automobiles onto the island, thus meaning that "[c]yclists, local buses, taxis, pedestrians and the occasional car of one of the island's inhabitants share the streets without any problems" (Peters 2006:128).

While cycling trips tend to decline with increasing age, 12 percent of Danes between ages 70 and 74 make such trips on a regular basis. Morabia and his colleagues (2010) conducted an experiment in which 18 people who commuted by car to Queens College in New York shifted to commuting for the next five days by public transportation. They found that "[c]ommuting by public transportation rather than by car increased energy expenditure (+124 kcal/day; P-.001) equivalent to the loss of 1 pound of body fat per 6 weeks"

(Morabia et al. 2010: 2388). In another study, Maizlish and his colleagues (2013) increased median daily walking and cycling from 4 to 22 minutes, reduced the burden of cardiovascular disease and diabetes by 14 percent, and decreased greenhouse gas emissions by 14 percent, but ironically increased the traffic injury burden by 39 percent because of inadequate protection from moving motor vehicles. Dutka (2013:138) reports that while the car-dependent United States has the highest obesity rate world-wide, Washington, DC, which is characterized by a high use of public transit, cycling, and walking trips exhibits the "third lowest obesity rate in the nation." Ironically, half of all driving trips in the United States are "within a 20-minute bicycle ride (3 miles or less), and a quarter of overall trips are within a 20-minute walk (1 mile or less)" (Mills 2013:178). Promotion of walking and cycling for short trips in American schools and media have the potential of diminishing the obesity epidemic in the United States, not to speak of other countries. In some cities, cyclists can transport their bicycles on trains, buses, or trams. For example, buses in largely car-dependent metropolitan Phoenix have a rack in the outside front that holds two or three bicycles. In Melbourne, cyclists can transport their bicycles on suburban trains and state-operated Victoria Line (V/L) trains that connect Melbourne with many outlying "country towns" in the state. However, it is important to note this practice is not a panacea because bicycles can take up much space that could potentially provide room for passengers. Although China constitutes a glaring example of increasing automobilization, the Chinese government has had various cities return cycle lanes that had previously been converted to motor vehicle lanes (Pendakur 2011:204).

While vastly more environmentally sustainable than automobiles or motorcycles, bicycles in great numbers require a lot of street space, consume a lot of space for parking, and potentially pose danger to other cyclists and pedestrians (Crawford 2002:175). Melbourne fortunately does have bicycle lanes on some streets but not all streets, thus placing the cyclist in greater danger as he or she seeks to navigate between moving traffic on the right-hand side and parked motor vehicles on the left-hand side. This particularly becomes problematic on streets with high moving traffic, thus forcing the cyclist to ride on the footpath, thus endangering or at least annoying pedestrians and generally breaking the law. In the carless city proposed by Crawford (2002:176), central boulevards are "one place where cyclists can ride fast" and where "bike lanes provide sufficient capacity for heavy bicycle traffic." Cyclists are quite capable of competing for spaces and in some extreme cases engage in "cycle rage" in which they attempt to cut each other off or become impatient with cyclists in front of them who are not moving as fast or agilely as they are (Lloyd 2017). Furthermore, while cyclists on mountain bikes are not competing for road space with motor vehicles, they are competing for space on trails with hikers and may even contribute to environmental damage, such as erosion to these trails (Vovanco 2013: 33).

While the resources that go into manufacturing bicycles are much less per passenger than those that go into manufacturing automobiles, motorcycles, and buses, nevertheless due to corporate globalization of the bicycle industry, many of them are now imported long distances from central manufacturing centers. Whereas between 1973 and 1974 of the 29 million bicycles sold in the United States, nearly 70 percent of them were manufactured by US companies, by 2007 an estimated 99.65 percent of the bicycles sold in the United States had been imported from China, Taiwan, and Hong Kong (Furness 2010:214). The process of shipping of bicycles over long distances results in greenhouse gas emissions that would not be incurred to nearly the same extent if they were manufactured locally.

Furness (2010:217) hints at the need for the need for a "more egalitarian vision of production that could simultaneously minimize the bicycle's environmental impact." I believe that ultimately such a vision can only be achieved within the parameters of a democratic eco-socialist world system in which there would be public or social ownership of the means of production, which would include the production of more environmentally sustainable modes of transportation, including bicycles which would be perhaps prioritized over other modes of transportation.

Other Measures: Telecommuting/Teleworking, Video Conferencing, Reviving Hitchhiking

Telecommunications advances now make it feasible to conduct a lot of work from home rather than an office in a business or government office complex. According to Black (2010:73), "telecommuting has already lessened the number of vehicles on the highway, although its other merits are questionable." For example, within a corporate structure, it might serve to further isolate workers from each other and mitigate against union organizing or the possibility of developing workers' democracy or at least a sense of community. Video conferencing or teleconferencing has often been proposed as a means for reducing travel, particularly by air, for business or other purposes, such as academic presentations. An often-downplayed dimension of conferences is that they constitute a perk which allows people to mix business with holidaying. Teleconferencing also makes the sort of informal social networking that occurs at conferences difficult to achieve.

Hitchhiking has been regarded with a great deal of ambivalence, perhaps particularly in the United States. During the 1930s and 1940s many Americans felt it was their civic duty to give needy people and later soldiers and sailors a lift. Also bohemians, such as the flappers in the 1920s and the hippies in the 1960s, viewed hitchhiking as a form of freedom (Packer 2008:19). After World War II, however, hitchhikers were often viewed as ominous and this perception probably intensified beginning in the 1970s. In a more trusting world, one committed to principles of social justice and environmental sustainability, however, the art of hitchhiking could be revived.

Autonomous Vehicles as a Possible Non-Reformist Reform

As I indicated earlier, I by no means regard autonomous vehicles, particularly privately owned autonomous automobiles, as a panacea to solving transportation problems. I suspect that they constitute the latest strategy on the part of automobile companies to perpetuate automobility in their pursuit of profits and to allow new ones such as Tesla and Google to position themselves in the still evolving domain of e-mobility. My biggest fear is that autonomous automobiles may detract from efforts to develop public transportation and displace trains, trolleys, and buses, just as conventional automobiles did in an earlier era. In certain situations, such as rural or remote areas, I can see a place for publicly owned autonomous vehicles that would connect people with small towns or train stations that would in turn connect them with cities. What place AVs may have in a more socially just and environmentally sustainable world remains to be seen. At any rate, I think that whatever AVs may exist should be a public good, one is that shared rather than one that is owned by a few privileged individuals or families, as is presently the case with electric automobiles.

Resisting the Culture of Consumption and Adopting Sustainable and Meaningful Consumption

Obviously, all humans need to consume a certain amount of food, clothing, and shelter to sustain themselves. Capitalism, however, converts "needs" into "wants" through voluminous and alluring advertisements and as a compensation for alienation in the workplace and everyday social life. From an eco-socialist perspective, Fred Magdoff and John Bellamy Foster (2011) argue that a democratic and egalitarian economic system will have to limit consumption levels to significantly less than they generally are for most middle-class people in developed societies. Australian eco-anarchist Ted Trainer (1998:8) argues that the "fundamental cause of the accelerating destruction of the global ecosystem is overproduction and overconsumption" of material goods, His notion of a "Simpler Way" incorporates the following principles: (1) far simpler material living standards; (2) high levels of self-sufficiency within households, nationally and especially within towns and neighborhoods; (3) relatively little long-distance trading and transportation; (4) small-scale economies in which most of the things we need are produced by local labor from local resources; (5) cooperative and participatory local systems; (6) an alternative economy that does not entail growth and requires far less work and production and consumption than the present one: (7) a commitment to human rights and social justice, particularly with regard to developing societies; and (8) a radically different culture (Trainer 2010).

Unfortunately, at least in developed societies, resistance to the culture of consumption remains confined to niche groups. Jonathan Neale (2008) warns climate activists not to talk about sacrifice to ordinary people. My comments of resisting the culture of consumption are directed primarily to the affluent, even the affluent in the working class, who turn to consumerism as a compensation for alienation in the workplace and in everyday life in developed societies. Most people in developed societies and the more affluent sectors in developing societies will need to scale back their consumption of material goods as well as restrict the number of holidays to far-away destinations that they take. This does not mean that people need to embrace asceticism but seek to live the good life in such a way that it is available to everyone and is environmentally sustainable and contributes to the creation of a safe climate.

While many people who claim to be concerned about the ecological crisis and climate change, getting them to restrict their driving or abandon it all together, let alone flying on a much more limited basis, is easier said than done. Boulder, situated at the base of the Colorado Rockies, constitutes an excellent example of this discrepancy. As John Tribbia (2007:238) observes, although Boulder has a reputation as being one of the "Top Green and Clean Cities" in the United States, there exists a "dissonance between what Boulderites think, believe, and politically support and what many actually do." Many Boulderites, despite the existence of a reasonably good public transportation system and network of walkways and cycle paths, still are wedded to driving around their rather sprawling city. Communities will need to mount public education campaigns, starting out in schools, which create a consciousness about the need for alternatives to automobiles as a mode of transportation.

Trainer (2015:22–22) delineates in detail practical steps that can be taken to achieve a Simpler Way; he includes a section on "Transport and Travel." Because most people would live close to work, the need for transportation, particularly by automobile would be greatly reduced. Trainer argues:

> A few cars, trucks and bulldozers would be needed. The vehicles in most use would be bicycles, with some but relatively little use of buses and trains. Horses could be used for some transport, especially carting goods the mostly short

distances required, for instance from local farms. They consume no oil, refuel themselves, reproduce themselves and do not need spare parts or expensive roads, but they do need the occasional vet. Most roads and freeways would be dug up and the space used for gardens. . . . Railway and bus production would be one of the few activities to take place in large centralised heavier industrial centres. (Trainer 2015:21–22)

Within the parameters of the Simpler Way, a few ships, large trucks, or airplanes would be needed to transport goods and people over long distances, partly due to a shortage of fuel but also to minimize greenhouse gas emissions. International travel would be rationed for special occasions, such as educational and cultural exchanges, but perhaps only once in a lifetime. Trainer (2015:22) maintains that "we could bring back wind ships, so you might study for your degree while on a leisurely trip around the world." Instead of people seeking adventure and relaxation in faraway exotic places, which they generally reach by airplane and then may explore in a hired automobile, they could seek these things closer to home, even in their own immediate surroundings.

Sustainable Trade

Over the past two centuries, global production has resulted in a tremendous cross-border trade of goods and services. While growth in international trade has been enhanced by free trade agreements and lower transportation costs, it relies heavily upon oil and contributes to greenhouse gas emissions in moving goods around the world by ship or airplane as well as trucks and trains. Furthermore, while developing countries, particularly China, are often criticized for their increasing greenhouse gas emissions, an appreciable amount of this is because developed countries are importing cheap resources and manufactured goods from developing countries. International aviation and marine fuels are exempt from international taxation schemes. The global food system has undergone a tremendous rise in "food miles"—a measurement of the distance that food travels from the site of production to the site of consumption. Vandana Shiva (2008) maintains that humanity can reduce food-miles by eating diverse, local, and fresh foods, rather than increasing greenhouse gas emissions through the spread of corporate industrial farming, nonlocal food supplies, and processed and packaged food. There is the need for the greening of shipping, which would rely upon solar and hydrogen energy-powered ships, sailing ships, and kite sails. Also, given that large quantities of products are now shipped by airplane and trucks, there is a strong need to revisit railroads and waterways as less energy-intensive modes of shipping.

CONCLUSION

The transitional steps that pertain to transportation and mobility that I have delineated, along with others which I have discussed elsewhere, constitute loose guidelines for shifting human societies or countries toward eco-socialism. I do not purport that my suggested guidelines are comprehensive because undoubtedly others could be added to the list. As humanity enters an era of increasingly dangerous climate change accompanied by tumultuous environmental and social consequences, it must consider alternatives that hopefully will circumvent dystopian scenarios that on-going socioeconomic, ecological, and climate crises face humanity if business continues as usual.

Sociologist Mimi Sheller (2018) suggests a new mobility paradigm, one that she terms *mobility justice*. Her framework seeks to replace the mobility injustices inherent in the capitalist world system, ranging from injustices related to class, gender, race, ethnicity, age, and disability to ones related to freedom of movement within countries and between countries and ones related to the circulation of resources, goods, energy, pollution, greenhouse gas emissions around the globe, in which both motor vehicles as well as airplanes, ships, and trains play a role. While space does not allow for a full exposition of Sheller's conception of mobility justice, while recognizing a certain need for people to move around the planet locally, regionally, nationally, and internationally, although less so than capitalist structures encourage them to do so, she asserts that the mobility justice entails the "protection of the planet itself through a living process of communing and the local mobilization of many networked mobile publics for the defense of the mobile commons" (Sheller 2018:171). In my view, eco-socialism is compatible with the basic principles of mobility justice, even though there may be subtle differences in their respective languages.

As noted earlier, eco-socialism rejects the capitalist treadmill of production and consumption and its associated growth model. Instead, it recognizes that humans live on an ecologically fragile planet with limited resources which must be sustained and renewed as much as possible for future generations. While presently or for the foreseeable future, the notion that democratic eco-socialism may be eventually be implemented in any society, developed or developing, or in various societies may appear absurd, history tells us that social changes can occur very quickly once social structural and environmental conditions have reached a tipping point.

As humanity moves forward into the twenty-first century, our survival as a species appears to be more and more precarious, particularly given that the impact of climate change in a multiplicity of ways looms on the horizon. I often hear climate activists in Australia say that we do not have enough time to transcend global capitalism to be able to create a safe climate for humanity. Thus, they argue that climate activists need to collaborate with more supposedly progressive corporate leaders and politicians in tackling the climate crisis within the parameters of the existing global political economy. In my view, combatting climate change and global capitalism need to be merged. While the more enlightened corporate elites and their political allies may permit some measures that contribute to climate change mitigation, they will certainly not consciously permit the eventual demise of global capitalism and the emergence of an eco-socialist world system. Green capitalism and existing climate regimes do not suffice to mitigate climate change in any serious vein. How can we expect the system that created the problem to solve the problem?

Envisioning alternatives to automobility is an important component in the struggle to transcend capitalism and replace it with a democratic eco-social world system. Conversely, shifting from automobiles to public transit—particularly intercity trains, suburban trains, and trams—would serve to diminish air pollution and greenhouse gas emissions, but these modes of transportation are not panaceas and they also can negatively impact settlement patterns, the environment, and health. Air travel also constitutes another form of mobility upon which I have briefly touched in this book and which needs much more consideration in terms of its social and environmental impacts than it has received to date. At any rate, shifting to a more sustainable transport system, which would entail far few private motor vehicles and far less reliance upon air transportation, will continue to be a vexed challenge even within a democratic eco-socialist world system. But I am convinced that a serious discussion about how to create a more socially just and environmentally sustainable transport system than presently exists within the parameters of the capitalist world system will require major social transformations at various levels, ranging from the local to the global.

References

Achoff, Nicole M. 2011. A tale of two crises: labour, capital and restructuring in the US auto industry. *In Socialist Register 2012—The Crisis and the Left.* Leo Panitch, Greg Albo, and Vivek Chibber, eds. Pp. 125–148. London: Merlin Press.

Adams, Rob. 2009. From industrial cities to eco-urbanity: The Melbourne case study. In *Eco-Urbanity: Towards Well-Mannered Built Environments*. Darko Radovic, ed. Pp. 33–46. London: Routledge.

Adey, Peter. 2009. *Aerial Life: Spaces, Mobilities, Affects. Chichester*, West Sussex: Wiley-Blackwell.

Agyeman, Julian. 2013. *Introducing Just Sustainabilities: Policy, Planning, and Practice.* London: Zed Books.

The Allen Consulting Group. 2013. The strategic role of the Australian manufacturing industry. Report to the Federal Chamber of Automotive Industries.

Aldred, Rachel. 2013. Incompetent or too competent? Negotiating everyday cycling identities in a motor dominated society. Mobilities 8:252–271.

Alvord, Katharine. 2000. *Divorce Your Car!: Ending the Love Affair with the Automobile.* Gabriola, BC: New Society Publishers.

Angus, Ian. 2016. *Facing the Anthropocene: Fossil Capitalism and the Crisis of the Earth System.* New York: Monthly Review Press.

Araghi, Yashar, Bert Van Wee, and Maarten Kroesen. 2017. Historic vehicles: An overview from a transport policy perspective. Transport Reviews 37: 571–589.

Archer, Neil. 2017. Genre on the road: The road movie as automobiles research. *Mobilities* 12: 509–519.

Arndt, Wulf-Holger, Xiaxu Bei, Guenter Emberger, Ulrich Fahl, Oliver Lah, Alexander Sohr, and Jan Tomaschek. 2014. Climate change and sustainable transportation in megacities. In *Mobility and Transportation: Concepts for Sustainable Transportation in the Future. Arndt*, Wulf-Holger ed. Pp. 12–22. Berlin: Jovis.

Arndt, Wulf-Holger and Norman Doege. 2014. Tehran-Karaj, Hashtgerd—integrated urban and transportation planning for GHG emission reduction. In *Mobility and Transportation: Concepts for Sustainable Transportation in the Future*. Arndt, Wulf-Holger ed. Pp. 59–76. Berlin: Jovis.

Association of American Railroads. 2008. Freight railroads & greenhouse gas emissions. June.

Assis, Uiara Wasconcelos de, and Glaucia Wasconcelos Silva. 2012. VLT: A sustainable solution to urban mobility, in Joao Pessoa-BP. *Work* 41:2169–2174.

Auer, Peter, Stephen Clibborn, and Russell D. Lansbury. 2012. Beyond our control: Labour adjustment in response to the global recession by multinational corporations in Australia. *ABL* 38:142–157.

Australian Automobile Association. 2016. *Transport affordability index*. Canberra: Australian Automobile Association.

Australian Bureau of Statistics. 2010. *Motor Vehicle Census*, Australia, Publication 9309.0, March.

———. 2013. 4102.0—Australian social trends—car nation, July. http://www.abs.gov.au/AUSSTATS/abs@.nsf/Lookup/4102.0Main+Features40July+2013, accessed December 26, 2017.

———. 2016. Method of travel to work. Cat. No. 2901.0.2016. Australian Capital Territory.

Australian Government. 2008. *Review of Australia's Automotive Industry: Final Report*. July 22.

Avila, Eric. 2014. *The Folklore of the Freeway: Race and Revolt in the Modernist City*. Minneapolis: University of Minnesota Press.

Baer, Hans A. 2018. *Democratic Eco-Socialism as a Real Utopia: Transitioning to an Alternative World System*. New York: Berghahn.

Baker, Stefan, Mark Zuidgeest, Heleen de Coninck, and Cornie Huizenga, 2014. Transport, development and climate change mitigation: Towards an integrated approach. *Transport Reviews* 34:335–355.

Balaker, Ted and Sam Staley. 2006. *The Road More Traveled: Why the Congestion Crisis Matters More Than You Think, and What We Can Do About It*. Lanham, MD Rowman & Littlefield.

Banister, David. 2005. *Unsustainable Transport: City Transport in the New Century*. London: Routledge.

Barbier, Edward B. 2010. *A Global Green New Deal: Rethinking the Economic Recovery*. Cambridge, UK: Cambridge University Press.

Bardou, Jean-Pierre, Jean-Jacques Chanaron, Patrick Fridenson, and James M. Laux (translated from French by James M. Laux. 1982. *The Automobile Revolution: The Impact of an Industry*. Chapel Hill: University of North Carolina Press.

Barker, Judith C. 1999. Road warriors: Driving behaviors on a Polynesian island. In *Anthropology in Public Health: Bridging Differences in Culture and Society*. Robert A. Hahn, ed. Pp. 211–234. New York: Oxford University Press.

Barme, Geremie R. 2002. Engines of the revolution: Car cultures in China. In *Autopia: Cars and Culture*. Peter Wollen and Joe Kerr, eds. Pp. 177–190. London: Reaktion Books.

Barnet, Richard and John Cavanaugh. 1994. *Global Dreams: Imperial Corporations in the New World Order*. New York: Simon & Schuster.

Batterbury, Simon. 2002. Cycling in Copenhagen, http://www.simonbatterbury.net.

Baudrillard, Jean. 1988. *America*, translated by Chris Turner. London: Verso.

Bayley, Stephen. 1986. *Sex, Drink and Fast Cars: The Creation and Consumption of Images*. London: Faber & Faber.
Beckmann, J. 2001. Risky mobility: The filtering of automobility's unintended consequences. PhD thesis, Sociology, University of Copenhagen.
Beder, Sharon. 2006. *Environmental Principles and Policies: An Interdisciplinary Introduction*. London: Earthscan.
Beed, Clive and Patrick Moriarty. 1988. Urban transport policy. *Journal of Australian Political Economy*. Number 22, pp. 57–68.
Behrens, Roger, Dorothy McCormick, Risper Orero, and Marilyn Omneh. 2017. Improving paratransit service: Lessons from inter-city *matatu* cooperatives in Kenya. *Transport Policy* 53: 79–88.
Behringer, Wolfgang. 2010. *A Cultural History of Climate*. Cambridge, UK: Polity.
Berners-Lee, Mike. 2010. *How Bad Are Bananas? The Carbon Footprint of Everything*. London: Profile Books.
Buecher, Ralph, John Pucher, Regine Gerike, and Thomas Goetschi. 2017. Reducing car dependence in the heart of Europe: Lessons from Germany, Austria, and Switzerland. *Transport Reviews* 37(1): 4–28.
Black, Anthony. 2009. Location, automobile policy, and multinational strategy: The position of South Africa in the global industry since 1995. *Growth and Change* 40:483–512.
Black, Edwin. 2006. *Internal Combustion: How Corporations and Governments Addicted the World to Oil and Derailed the Alternatives*. New York: St. Martin's Press.
Black, William R. 2010. *Sustainable Transportation: Problems and Solutions*. New York: Guilford Press.
Blanke, David. 2007. *Hell on Wheels: The Promise and Peril of America's Car Culture, 1900–1940*. Lawrence: University Press of Kansas.
Blow L. and I. Crawford. 1997. The distributional effects of taxes on private motoring London: Institute of Fiscal Studies. December.
BMC-Leyland Australia Heritage Group. 2012. *Building Cars in Australia*. Sydney: Halstead Press.
Boehm, Steffan, Campbell Jones, Chris Land, and Mat Paterson. 2006. Introduction: impossibilities of automobility. *Sociological Review*. Supplement 1.
Bohren, Lenora. 2009. Car culture and decision-making: Choice and climate change. In *Anthropology & Climate Change: From Encounters to Actions*, eds. Pp. 370–379. Walnut Creek, CA: Left Coast Press.
Bongardt, Daniel, Felix Creutzig, Hanna Hueging, Ko Sakamoto, Stefan Bakker, Sudhir Gota, and Susanne Boehler-Baedeker. 2013. *Low-Carbon Land Transport: Policy Handbook*. London: Earthscan Routledge.
Bond, Patrick. 2015. BRICS and the sub-imperial location. In *BRICS: An Anti-Capitalist Critique* Patrick Bond and Ana Garcia, eds. Pp. 15–26. London: Pluto Press.
Bonneuil, Christophe and Jean-Baptiste Fressoz. 2016. *The Shock of the Anthropocene*. London: Verso.
Boswell, Terry and Christopher Chase-Dunn. 2000. *The Spiral of Capitalism and Socialism*. Boulder, CO: Lynne Rienner.
Bowdon, Rob. 2004. *Transportation: Our Impact on the Planet*. Lewis, East Sussex: White-Thomson Publishing.

Bradsher, Keith. 2002. *High and Mighty: SUVs-the World's Most Dangerous Vehicles and How They Got That Way.* New York: PublicAffairs.

Brand, Christian and John M. Preston. 2010. "60-20 emission"—The unequal distribution of greenhouse gas emissions from personal, non-business travel in the UK. *Transport Policy* 17: 9–19.

Bridger, Rose. 2013. *Plane Truth: Aviation's Real Impact on People and the Environment.* London: Pluto Press.

Bright, Brenda. 1998. "Heart like a car": Hispano/Chicano culture in northern New Mexico. *American Ethnologist* 25:583–609.

Brottman, Mikita. 2001. Introduction. In *Car Crash Culture.* Mikita Brottman, ed. Pp. xi–xliii). New York: Palgrave.

Brown, Anne E., Evelyn Blumberg, Brian D. Taylor, Kelcie Ralph, Edward J. Bloustein, and Carole Turley Voulgaris. 2016. A taste for transit? Analyzing public transit use trends among youth. *Journal of Public Transportation* 19(1):49–67.

Brown, Lester. 2001. Pavement is replacing the world's croplands. Grist, March 1, http://grist.org/article/rice/.

Broz, Ludek and Joachim Otto Habeck. 2015. Siberian automobility: From the joy of destination to the joy of driving there. *Mobilities* 10:552–570.

Bruder, Jessica. 2017. *Nomadland: Surviving America in the Twenty-First Century.* New York: W.W. Norton & Company.

Buechler, Ralph, John Pucher, Regine Gerike, and Thomas Goetschi. 2017. Reducing car dependence in the heart of Europe: Lessons from Germany, Austria, and Switzerland. *Transport Reviews* 37(1):4–28.

Bull, Michael. 2001. Soundscapes of the car: A critical study of automobile habitation. In *Car Cultures.* Daniel Bell, ed. Pp. 185–202. London: Berg.

Bunzhez, Corina Larisa. 2015. The future of transportation—autonomous vehicles. *International Journal of Economic Practices and Theories* 5: 447–454.

Business Monitor International. 2009. Business environment ratings—Russia auto reports Q2, pp. 15–18.

Cahill, Michael. 2010. *Transport, Environment and Society.* Maidenhead, Berkshire, UK: Open University Press.

Callenbach, Ernest. 1975. *Ecotopia: The Notebooks and Reports of William Weston.* New York: Bantam Books.

Calthorpe, Peter. 2016. Urbanism and global sprawl. In *Can a City Be Sustainable?* Lisa Mastny, ed. Pp. 91–108. Washington, DC: Island Press.

Carpenter, T. G. 1994. *The Environmental Impact of Railways.* New York: Wiley. http://siteresources.worldbank.org/EXTRAILWAYS/Resources/515244-1268663980770/greenhouse.pdf.

Carrabine, Eamonn and Brian Longhurst. 2002. Consuming the car: Anticipation, use and meaning in contemporary youth culture. *Sociological Review* 50(2):181–196.

Centre for International Economics. 2016. Reducing greenhouse gas emissions for light vehicles: Compulsory standards and other policy options (prepared for the Australian Automobile Association), August, http//:www.TheCIE.com.au.

Cevero, Robert. 2013. Linking urban transport and land use in developing countries. *Journal of Transport and Land Use* 6(1): 7–24.

Chafe, Zoe. 2008. Air travel reaches new heights. In *Vital Signs 2007–2008: The Trends That Are Shaping Our Future*. Linda Starke, ed. Pp. 70–71. New York: W.W. Norton & Company.

Chan, Anita, Yiu Por Chen, Yuhua Xie, Zhao Wei, and Cathy Walker/ 2014. Disposable bodies and labor rights: Workers in China's automotive industry. *Journal of Labor & History* 17:509–529.

Chelcea, Liviu and Ioana Iancu. 2015. An anthropology of parking: Infrastructures of automobility, work, and circulation. *Anthropology of Work Review* 36(2):62–73.

Chakraborty, Jayajit, 2009. Automobiles, air toxics, and adverse health risks: Environmental inequities in Tampa Bay, Florida. *Annals of the Association of American Geographers* 99:674–697.

Chomsky, Noam. 2012. *Occupy*. Brooklyn: Zuccotti Park Press.

Choudhary, Arti and Sharad Gokhale. 2016. Urban real-world driving traffic emissions during interruption and congestion. *Transportation Research* Part D 43:59–70.

Clark, Duncan. 2009. *The Rough Guide to Green Living*. London: Rough Guides.

Clarsen, Georgine. 2017. "Australia—drive it like you stole it": Automobility as a medium of communication in settler Australia. *Mobilities* 12: 520–533.

Clibborn, Stephen, Russell D. Lansbury, and Chris F. Wright. 2016. Who killed the Australian automotive industry: The employers, government or trade unions? *Economic Papers* 15(1):2–15.

Coffman, Makena, Paul Bernstein, and Sherilyn Wee. 2017. Electric vehicles revisited: A review of factors that affect adoption. *Transport Reviews* 37(1) 79–93.

Cohen, Maurie J. 2006. A social problems framework for the critical appraisal of automobility and sustainable system innovation. *Mobilities* 1:23–38.

Collin-Lange, Virgile. 2013. Sociabilities in motion: Automobility and car cruising in Iceland. *Mobilities* 8:406–423.

Conley, Tom. 2009. *The Vulnerable Country: Australia and the Global Economy*. Sydney: UNSW Press.

Conlon, Robert and John Perkins. 2001. *Wheels and Deals: The Automobile Industry in Twentieth-Century Australia*. Aldershot, UK: Ashgate.

Cooper, Jai and Terry Leahy. 2017. Cycletopia in the sticks: Bicycle advocacy beyond the city limits. *Mobilities* 12:611–627.

Coopley, Richard. 2010. Power without knowledge? Foucault and Fordism, c. 1900–50. *Labor History* 51:107–125.

Crawford, J.H. 2002. *Carfree Cities: When Streets Are for People, Everyone Wins*. Utrecht: International Books.

Cwerner, Saulo B. 2006. Vertical flight and urban mobilities: The promise and reality of helicopter travel. *Mobilities* 1:191–215.

Curtis, Carey and Nicholas Low. 2012. *Institutional Barriers to Sustainable Transport*. Farnham, Surrey: Ashgate.

Dant, Tim. 2005. The driver-car. In *Automobilities*. Mike Featherstone, Nigel Thrift, and John Urry, eds. Pp. 61–79. London: Sage.

Dalakoglou, Dimitris. 2010. The road: An ethnography of the Albanian-Greek cross-border motorway. *American Ethnologist* 37:132–149.

———. 2012. "The road from capitalism to capitalism": Infrastructures of (post)socialism in Albania. *Mobilities* 7:571–586.
Daly, Herman and John B. Cobb, Jr. 1990. *For the Common Good: Redirecting the Economy Toward Community, the Environment, and a Sustainable Future*. Boston: Beacon Press.
Damoense-Azevedo, and Andre C. Jordann. 2011. Trade patterns in the automobile industry: Some evidence from South Africa. *Journal of African Business* 12:154–177.
Dauvergne, Peter. 2008. *The Shadows of Consumption: Consequence for the Global Environment*. Cambridge, MA: MIT Press.
Davis, Mark. 2008. *The Land of Plenty: Australia in the 2000s*. Melbourne: Melbourne University Press.
Davis, Tony. 2016. Revolution on wheels: Will Australia be participant or spectator? *Griffith Review*, No. 52, Pp. 179–201.
Davison, Geoffrey and Ang Wei Ping. 2016. City View: Singapore. In *Can a City Be Sustainable?* Lisa Mastny, Ed. Pp. 211–216. Washington, DC: Island Press.
Davison, Graeme. 2004. *Car Wars: How the Car Won Our Hearts and Conquered Our Cities*. Sydney: Allen & Unwin.
———. 2016. *City Dreamers: The Urban Imagination in Australia*. Sydney: NewSouth Publishers.
Dawson, Andrew. 2017. Driven to sanity: An ethnographic critique of the senses in automobilities. *The Australian Journal of Anthropology* 28:3–20.
De Vlieger, I., D. Keukeleere, and J. Kretzschmar. 2000. Environmental effects of driving behaviour and congestion related to passenger cars. *Atmospheric Environment* 34:4649–4655.
Delbosc, Alexa and Graham Currie. 2013. Causes of youth licensing decline: A synthesis of evidence. *Transport Reviews* 33:271–290.
Dennis, Kingsley and John Urry. 2009. *After the Car*. London: Polity.
Dery, Mark. 2006. "Always crashing in the same car": A head-on collision with the technosphere. *Sociological Review* 54(1): 223–239. Supplement.
Desmond, Jane. 2013. Requiem for roadkill: Death and denial on America's roads. In *Environmental Anthropology: Future Directions*. Helen Kopnina and Eleanore Shoreman-Ouimet, eds. Pp. 46–58. London: Routledge.
Dettelbach, Cynthia Golomb. 1976. *In the Driver's Seat: The Automobile in American Literature and Popular Culture*. Westport, CT: Greenwood Press.
Diamond, Jared. 2005. *Collapse: How Societies Choose to Fail or Succeed*. New York: Penguin.
Dicken, Peter. 2003. *Global Shift: Reshaping the Global Economic Map in the 21st Century* (4th edition). New York: Guilford Press.
Dixit, Vinayak V., Sai Chand, and Divya J. Nair. 2016. Autonomous vehicles: Disengagements, accidents and reaction times. *PLOS*, 11(12), December 20, Pp. 1–14.
Dobbs, Lynn. 2005. Wedded to the car: women, employment and the importance of private transport. *Transport Policy* 12: 266–278.
Dodson, Jago and Neil Sipe. 2008. *Shocking the Suburbs: Oil Vulnerability in the Australian City*. Sydney: UNSW Press.

Dowling, Robyn and Jennifer Kent. 2015. Practice and public-private partnerships in sustainable transport governance: The case of car sharing in Sydney, Australia. *Transport Policy* 40: 58–64.
Doyle, Jack. 2000. *Taken for a Ride: Detroit's Big Three and the Politics of Air Pollution*. New York: Four Walls Eight Windows.
DuBoff, Richard B. 1989. *Accumulation and Power: An Economic History of the United States*. Armonk, NY: M.E. Sharpe.
Duffy, Enda. 2009. *The Speed Handbook: Velocity, Pleasure, Modernism*. Durham, NC: Duke University Press.
Dunn, James A. 1998. *Driving Forces: The Automobile, Its Enemies and the Politics of Mobility*. Washington, DC: Brookings Institution.
———. 1999. The politics of automobility. *The Brookings Review* 17(1):15.
Durden, Tyler. 2014. Where the world's unsold cars go to die. May 5, website.
Dutka, Projjal. 2013. Taking the car out of the carbon: Mass transit and emission avoidance. In *Transport Beyond Oil: Policy Choices for a Multimodal Future*. John L. Renne and Billy Fields, ed. Pp. 126–140. Washington, DC: Island Press.
Eckersley, Robyn. 2004. *The Green State: Rethinking Democracy and Sovereignty*. Cambridge, MA: MIT Press.
Edensor, Tim. 2005. Automobility and national identity: Representation, geography and driving practice. *Automobilities*. Mike Featherstone, Nigel Thrift, and John Urry, eds. Pp. 101–120. London: Sage.
Emanuel, Kamala. 2016. Inside every princess . . . *Green Left Weekly*, August 2, p. 7.
Emberger, Guenter. 2014. Ho Ch Minh City, Vietnam—can HCMC research its CO_2 targets in the transport sector? In *Mobility and Transportation: Concepts for Sustainable Transportation in the Future*. Arndt, Wulf-Holger ed. Pp. 25–37. Berlin: Jovis.
Environmental Defense Fund. 2007. Cars by the numbers: statistics on automobiles and their global warming. https://www.edf.org/sites/default/files/5301_Globalwarmingontheroad_0.pdf.
Eriksen, Thomas Hylland. 2016. Overheating: An Anthropology of *Accelerated Change*. London: Pluto Press.
Etyemezian, V. et al. 2005. Results from a pilot-scale air quality study in Addis Ababa, Ethiopia. *Atmospheric Environment* 39:7849–7860.
European Conference of Ministers of Transport 2007. Cutting transport CO_2 emissions: What progress?—summary document.
Evans, Antony. 2013. The rebound effect in the aviation sector. *Energy Economics* 36: 168–165.
Evans, Antony and Andreas Schaefer. 2013. The rebound effect in the aviation sector. *Energy Economics* 36:158–165.
Evans, Leonard. 2006. The dramatic failure of U.S. traffic safety policy. *TR [Transportation Research] News*, No. 242, 28–31.
Ewing, Jack. 2017. *Faster, Higher, Farther: The Inside Story of the Volkswagen Scandal*. London: Bantam Press.
Fallon, I. and D. O'Neill. 2005. The world's first automobile fatality. *Accident Analysis and Prevention* 37:601–603.

Fargione, Joesph E., Richard J. Plevin, and Jason D. Hill. 2010. The ecological impact of biofuels. *Annual Review of Ecological Evolutionary Systems* 41:351–377.

Fava, Valentina. 2011. The elusive people's car: Imagined automobility and productive practices along the "Czechoslovak road to socialism" (1945–1968). In *The Socialist Car: Automobility in the Eastern Bloc*. Lewis H. Siegelbaum, ed. Pp. 17–29. Ithaca, NY: Cornell University Press.

Fava, Valentina and Luminta Gatejel. 2017. East-West cooperation in the automotive industry: Enterprises, mobility, production. *Journal of Transport History* 38(1): 11–19.

Fellman, Gordon. 1986. Neighborhood protest of an urban highway. In *Transport Sociology: Social Aspects of Transport Planning*. Enne de Boer, ed. Pp. 29–38. Oxford, UK: Pergamon Press.

Feng, Jianxi, Martin Dijst, Bart Wissink, and Jan Prillwitz. 2017. Changing travel behaviour in urban China: Evidence from Nanjing 2008–2011. *Transport Policy* 53: 1–10.

Flachsbart, Peter G. 1992. Human exposure to motor vehicle air pollution. In *Motor Vehicle Air Pollution: Public Health Impact and Control Measures*. David Mage and Olivier Zali, eds. Pp. 85–106. Geneva: World Health Organization and Geneva Department of Public Health.

Fleming, Grant, David Merrett, and Simon Ville. 2004. *The Big End of Town: Big Business and Corporate Leadership in Twentieth Century Australia*. Melbourne: Cambridge University Press.

Flink, James J. 1973. *The Car Culture*. Cambridge, MA: MIT Press.

———. 1988. *The Automobile Age*. Cambridge, MA: MIT Press.

Flonneau, Mathieu. 2006. City infrastructures and city dwellers: Accommodating the automobile in twentieth-century Paris. *Journal of Transport History* 27(1): 93–114.

Ford, Henry. 1922. *Ford Ideals: Being a Selection from "Mr. Ford's Page"* in the Dearborn Independent. Dearborn, MI: Dearborn.

Forest, James J.F. and Matthew V. Sousa. 2006. *Oil and Terrorism in the New Gulf: Framing U.S. Energy and Security Policies for the Gulf of Guinea*. Lanham, MD: Lexington Books.

Forman, Jason L., Aileen Y. Watchko, and Maria Sequi-Gomez. 2011. Death and injury from automobile collisions: An overlooked epidemic. Medical Anthropology 30:241–246.

Fortune 2016. Here are the top 10 corporations in Fortune 500. http://www.fortune.com/2016/06/06/fortune-500-top-10-companies.

Foster, John Bellamy. 2000. *Marx's Ecology*. New York: Monthly Review Press.

———. 2002. *Ecology Against Capitalism*. New York: Monthly Review Press.

———. 2009. *The Ecological Revolution*. New York: Monthly Review Press.

Foster, Mark S. 2003. *A Nation on Wheels: The Automobile Culture in America since 1945*. Belmont, CA: Thomson Learning.

Frank, Harmut. 1986. Mass transport and class struggle. In *Transport Sociology: Social Aspects of Transport Planning*. Enne de Boer, ed. Pp. 211–222. Oxford, UK: Pergamon Press.

Freund, Peter S. and George Martin. 1993. *The Ecology of the Automobile*. Montreal: Black Rose Books.

———. 2007. Hyperautomobility, the social organization of space, and health. *Mobilities* 2(1):37–49,
———. 2008. Fast cars/fast foods: Hyperconsumption and its health and environmental consequences. *Social Theory & Health* 6:309–322.
Freudendal-Pedersen, Malene. 2009. *Mobility in Daily Life: Between Freedom and Unfreedom.* Farnham, Surrey: Ashgate.
Furness, Zack. 2007. Critical Mass, urban space and velomobility. *Mobilities* 2:299–319.
———. 2010. *One Less Car: Bicycling and the Politics of Automobility.* Philadelphia: Temple University Press.
Garrison, William L. and David M. Levinson. 2014. *The Transportation Experience* (2nd edition). Oxford UP.
Garrison, William L. and David Levinson. 2006. *The Transportation Experience: Policy, Planning, and Deployment.* Oxford, UK: Oxford University Press.
———. 2014. The *Transportation Experience* (2nd edition). Oxford, UK: Oxford University Press.
Gartman, David. 1994. *Auto Opium: A Social History of the American Automobile Design.* London: Routledge.
Garvey, Pauline. 2001. Driving, drinking and dating in Norway. In *Car Cultures.* Daniel Miller, ed. Pp. 133–152. London: Berg.
Gatejel, Luminita. 2016. Appealing for a car: Consumption policies and entitlement in the USSR, the GDR, and Romaniam 1950s-1980s. *Slavic Review* 75(1); 122–145.
Gautier, Catherine. 2008. *Oil, Water, and Climate: An Introduction.* Cambridge, UK: Cambridge University Press.
Gelbspan, Ross. 2004. *Burning Point: How Politicians, Big Oil and Coal, Journalists, and Activists Are Fueling the Climate Crisis—and What We Can Do to Avert Disaster.* New York: Basic Books.
Gelder, Ken. 1995. *Mad Max* and Aboriginal automation: Putting cars to use in contemporary Australian films and narratives. In *The Motor Car and the Popular Culture in the 20th Century.* David Thomas, Len Holden, and Tim Claydon, eds. Pp. 56–66. Aldershot, UK: Ashgate.
George, Rose. 2013. *Ninety Percent of Everything: Inside Shipping, the Invisible Industry That Puts Clothes on Your Back, Gas in Your Car, and Food on Your Table.* New York: Metropolitan Books.
Gewald, Jan-Bart, Sabine Luning, and Klaas van Walraven. 2009. Motor vehicles and people in Africa: An introduction. In *The Speed of Change: Motor Vehicles and People in Africa, 1890-2000.* Jan-Bart Gewald, Sabine Luning, and Klass van Walraven, eds Pp. 1-18. Leiden, the Netherlands: Brill.
Gilroy, Paul. 2001. Driving while black. In *Car Cultures,* Daniel Miller, ed. Pp. 81–104. Oxford, UK: Berg.
Gilbert, Richard and Anthony Perl. 2010. *Transport Revolutions: Moving People and Freight without Oil.* London: Earthscan.
Godard, Xavier. 2011. Poverty and urban mobility: Diagnosis toward a new understanding. In *Urban Transport in the Developing World: A Handbook of Policy and Practice.* Harry T. Dimitrious and Ralph Gakenheimer, eds. Pp. 232–261. Cheltenham, UK: Edward Elgar.

Goddard, Stephen B. 1994. *Getting There: The Epic Struggle between Road and Rail in the American Century.* Chicago: University of Chicago Press.

Golten, Robert J., Oliver A. Houck, and Richard Munson, eds. 1977. *The End of the Road: A Citizen's Guide to Transportation Problem Solving.* Washington, DC: National Wildlife Federation.

Goodall, Chris. 2007. *How to Live a Low-Carbon Life: The Individual's Guide to Stopping Climate Change.* London: Earthscan.

Goodman, Randall H. 2012. The future of carfree development in York, UK. In Special edition: a future beyond the car? Steve Maila, ed. Pp. 41–64. *World Transport, Policy & Practice* 17(4).

Goodwin, Phil and Kurt van Dender. 2013. "Peak car"—themes and issues. *Transport Review* 33:243–254.

Gordon, Deborah and David Burwell. 2013. The role of transportation in climate disruption. In *Transport Beyond Oil: Policy Choices for a Multimodal Future.* John L. Renne and Billy Fields, eds. Pp. 11–30. Washington, DC: Island Press.

Gotschi, Thomas, Jan Garrard, and Billie Giles-Gorti. 2016. Cycling as a part of daily life: A review of health perspectives. *Transport Reviews* 36:45–71.

Graves-Brown, Paul. 1997. From highway to superhighway: The sustainability, symbolism and situated practices of car culture. *Social Analysis* 41(1): 64–75.

Gordon, Sarah H. 1996. *Passage to Union: How the Railroads Transformed American Life, 1829–1929.* Chicago: Ivan R. Dee.

Gough, Kathleen. 1968. Anthropology and imperialism. *Monthly Review*, April, Pp. 12–27.

Gray, Ian. 2011. Maintaining the power of the central governments: Regional land transport in the Australian Federation, 1850-2007. *Journal of Transport History.* 30(1):22–39.

Greer, Ian. 2008. Organised industrial relations in the information economy: The German automotive sectors as a test case. *New Technology, Work and Employment* 23:181–196.

Groenhart, Lucy and Paul Mees. 2014. The journey to work. In Gleeson, Brendan and Beau B. Beza. 2014. *The Public City: Essays in Honour of Paul Mees.* Brendan Gleeson and Beau B. Beza, eds. Pp. 121–131. Melbourne: Melbourne University Press.

Glover, Leigh. 2011. GAMUT papers: Personal carbon budgets for transport. GAMUT Australasian Centre for the Governance and Management of Urban Transport, University of Melbourne.

Goessling, Stefan. 2017. *The Psychology of the Car.* Amsterdam: Elsevier Inc.

Goessling, Stefan and Daniel Metzler. 2017. Germany's climate policy: Facing an automobile dilemma. *Energy Policy* 105:418–428.

Goodall, Chris. 2007. *How to Live a Low-Carbon Life: The Individual's Guide to Stopping Climate Change.* London: Earthscan.

Gorz, Andre. 1973. *Socialism and Revolution.* Garden City, NY: Anchor.

———. 1973. The social ideology of the motorcar. Originally published in *Le Sauvage*, September-October.

Haas, Tobias and Hendrik Sander. 2016. Shortcomings and perspectives of the German *Energiewende*. *Socialism and Democracy* 30(2):121–143.

Hagman, Olle. 2006. Morning queues and parking problems: On the broken promises of the automobile. *Mobilities* 1:63–74.

———. 2010. Driving pleasure: A key concept in Swedish car culture. *Mobilities* 5:25–39.

Haigh, Gideon. 2013. *End of the Road?* Melbourne: Penguin Books.

Halder, Dilip. 2006. *Urban Transport in India: Crisis and Cure.* New Delhi: Bookwell.

Hamer, Mick. 1987. *Wheels within Wheels: A Study of the Road Lobby.* London: Routledge & Kegan Paul.

Hansen, Arve. 2017. Hanoi on wheels: Emerging automobility in the land of the motorbike. *Mobilities* 12:628–645.

Harari, Yuval Noah. 2015. *Homo Deus: A Brief History of Tomorrow.* London: Harvell Seeker.

Harding, Simon E., Madhav G. Badami, Conor C.O. Reynolds, and Milind, Kandlikar. 2016. Auto-rickshaws in Indian cities: Public perceptions and operational realities. *Transport Policy* 52: 143–152.

Harper, Corey D., Chris T. Hendrickson, Sonia Mangones, and Constantine Samaras. 2016. Estimating potential increases in travel with autonomous vehicles for the non-driving, elderly and people with travel-restrictive medical conditions. *Transportation Research Part C* 72: 1–9.

Hart, Jennifer. 2016. *"Sweet Not Always": Automobility, State Power, and the Politics of Development, 1980s–1990s.* Bloomington: Indiana University Press.

Harvey, David. 1990. *The Condition of Postmodernity.* Cambridge, UK: Blackwell.

———. 2014. Cities or urbanization. In *Implosion/Explosions: Towards a Study of Planetary Urbanization?* Neil Brenner, ed. Pp. 52–66. Berlin: Jovis Verlag.

Harvey, Penny and Hannah Knox. 2012. The enchantments of infrastructure. *Mobilities* 7:521–546.

Hawkins, Jason and Khandker Nurul Habib. 2019. Integrated models of land use and transportation for the autonomous vehicle revolution. *Transport Reviews* 39:66–83.

Heinberg, Richard. 2015. *Afterburn: Society beyond Fossil Fuels.* Gabriola Island, BC: New Society Publishers.

Heinberg, Richard and David Fridley. 2016. *Our Renewable Future: Laying the Path for Clean Energy.* Post Carbon Institute.

Henriksson, Lars. 2013. Cars, crisis, climate change and class struggle. In *Trade Unions in the Green Economy: Working for the Environment.* Nora Raethzel and David Uzell, eds. Pp. 78–86. London: Routledge.

Huang, Kai, Pavlos Kanaroglou, and Xiaozhou Zhang. 2016. *Transport Research Part D* 49:1–17.

Hickman, Robin and David Banister. 2014. *Transport, Climate Change and the City.* London: Routledge.

Hill, Steven. 2015. *Raw Deal: How the "Uber Economy" and Runaway Capitalism Are Screwing American Workers.* NY: St Martin's Press.

Hillman, Mayer. 2012. The implications of climate change for the future of the car. In Special edition: A future beyond the car? Steve Meila, ed. Pp. 18–29. *World Transport, Policy & Practice* 17(4):18–29.

Hirschman, Elizabeth C. 2003. Men, dogs, guns and cars: The semiotics of rugged individualism. *Journal of Advertising* 32(1): 9–22.

Hoffler, Don. 2012. *Holden Days: From the Original 48–215 'FX' to the 1966 HR*. Kent Town, South Australia: Wakefield Press.

Holland, Brian and Juan Wei. 2016. City view: Portland, Oregon, United States. In *Can a City Be Sustainable?* Lisa Mastny, ed. Pp. 291–296. Washington, DC: Island Press.

Hopkins, Debbie. 2016. Can environmental awareness explain declining preference for car-based mobility amongst generation Y? A qualitative examination of learn to drive behaviours. *Transportation Research Part A* 94:149–163.

Hoyle, Brian and Richard Knowles. 1998. *Modern Transport Geography* (2nd edition). Chichester, UK: John Wiley & Sons.

Huijbens, Edward H. and Karl Benediktsson. 2007. Practising highland hetorotopias: Automobility in the interior of Iceland. *Mobilities* 2:143–165.

Hutton, Barry. 2013. *Planning Sustainable Transport*. London: Earthscan from Routledge.

IATA. 2013a. Air passenger market analysis. December, http//:www.iata.org/carrier-tracker. International Association of Public Transport. N.d. Public transport and CO_2 emissions. Website.

———. 2013b. Figures reveal demand for air travel. https://www.breakingtravelnews.com/news/article/iata-reports-demand-for-air-travel-stays-firm-but-with-regional-variations/.

———. 2014. Transport action plan (provisional copy). Submitted to Climate Summit 2014, New York, September 23.

IEA/OECD. 2013. Renewable energy in transport. https://www.iea.org/media/training/presentations/Day_2_Renewables_5_Transport.pdf.

International Association of Public Transport. 2014. UTIP Declaration on Climate. http://slocat.net/international-association-of-public-transport-initiative.

International Energy Agency. 2015. World Energy Outlook 2015. https://www.iea.org/publications/freepublications/publication/WEO.

International Maritime Organization. 2008. Updated Study on Greenhouse Gas Emissions from Ships. September 1.

International Organization of Motor Vehicle Manufacturers (OICA). 2014 Production Statistics. http://www.oica.net/category/production-statistics/2014-statistics/.

International Transport Forum. 2010. Reducing Transport Greenhouse Gas Emissions: Trends & Data. OECD. https://www.itf-oecd.org/sites/default/files/docs/10ghgtrends.pdf.

Jackson, Laura E. 2003. The relationship of urban design to human health and condition. *Landscape and Urban Planning* 64: 191–200.

Jacobs, Sherelle. 2014. Africa on wheels—the continent's auto market. *African Business*, April, Pp. 12–23.

Jegede, Abayomi. 2019. Top 10 most polluted cities in India. *The Daily Records*, January 2, http://www.thedailyrecords.com/2018-2019-2020-2021/world-famous-top-10-list/world/most-polluted-cities-india-least/8542/.

Jeske, Christine. 2016. Are cars the new cows? Changing wealth goods and moral economies in South Africa. *American Anthropologist* 118:433–494.

Josa, Sergio Oliete and Francesc Magrinya. 2018. Patchwork in an interconnected world: The challenges of transport networks in Sub-Saharan Africa. *Transport Reviews* 38:710–736.

Kahn Ribeiro, S., Kobayashi, S., Beuthe, M., Gasca, J., Greene, D., Lee, D. S., et al. (2007). Transport and its infrastructure. In *Climate change 2007: Mitigation.* B. Metz, O. R. Davidson, P. R. Bosch, R. Dave, and L. A. Meyer, eds. Contribution of Working Group III to the fourth assessment report of the Intergovernmental Panel on Climate Change. Cambridge: Cambridge University Press.

Kasarda, John D. and Greg Lindsay. 2011. *Aerotropolis: The Way We'll Live Next.* New York: Farrar, Straus and Giroux.

Kay, Jane Holtz. 1997. *Asphalt Nation: How the Automobile Took Over America and How We Can Take It Back.* New York: Crown.

Keller, Maryann. 1993. *Collision: GM, Toyota, Volkswagen and the Race to Own the 21st Century.* New York: Currency Doubleday.

Kelly, Jane-Frances and Paul Donegan. 2015. City Limits: *Why Australia's Cities Are Broken and How We Can Fix Them.* Melbourne: Melbourne University Press.

Kelsey, J.L. et al. 1984. Acute prolapsed lumbar intervertebral disc: An epidemiologic study with special reference to driving automobiles and cigarette smoking. *Spine* 9:608–613.

Kelsey, J.L., A.L. Golden, and D.J. Mundt. 1990. Low back pain/prolapsed lumbar intervertebral disc. *Rheumatic Disease Clinics of North America* 16(3):699–715.

Kelsey, J. L. and R.J. Hardy. 1975. Driving of motor vehicles as a risk factor for acute herniated lumbar intervertebral disc. *American Journal of Epidemiology* 102:62–73.

Kemp, R. 1995. The European high speed network. In *Passenger Transport after 2000 AD.* G.B.R. Feilden, A.H. Wickens and I.R. Yates, eds. Pp. 64-84. London: E and FN Spon for the Royal Society.

Kent, Jennifer L. 2015. Still feeling the car—the role of comfort in sustaining private car use. *Mobilities* 10:726–747.

Kenworthy, Jeffrey. 2011. An international comparative perspective on fast-rising motorization and automobile dependence. In *Urban Transport in the Developing World: A Handbook of Policy and Practice.* Harry T. and Ralph Gakenheimer, eds. Pp. 71–112. Cheltenham, UK: Edward Elgar.

Keucheyan, Ramzig. 2016. *Nature is a Battlefield.* London: Polity.

Kingsley, Dennis and John Urry. 2009. *After the Car.* London: Polity.

Klare, Michael. 2008. *Rising Powers, Shrinking Planet: The New Geopolitics of Energy.* New York: Metropolitan Books.

Klein, Nicholas J. and Michael J. Smart. 2017. Millennials and car ownership: Less money, fewer cars. *Transport Policy* 53: 20–29.

Kline, Ronald and Trevor Pinch. 1996. Users as agents of technological change: The social construction of the automobile in rural United States. *Technology and Culture* 37: 763–795.

Konczal, Mike. 2014. Socialize Uber: It's easier than you think. *The Nation*, December 10, https://www.thenation.com/article/socialize-uber.

Koch, Max. 2012. *Capitalism and Climate Change: Theoretical Discussion, Historical Development and Policy Responses.* NY: Palgrave Macmillan.

Korten, David. 2001. *When Corporations Rule the World* (2nd edition). San Francisco: Kumerian Press.

Koshar, Rudy. 2005. Cars and nations: Anglo-German perspectives on automobility between the world wars. *Automobilities*. Mike Featherstone, Nigel Thrift, and John Urry, eds. Pp. 121–144. London: Sage.

Koslowsky, Meni, Abraham N. Kluger, and Mordechai Reich. 1995. *Commuting Stress: Causes, Effects, and Methods of Coping*. New York: Plenum Press.

Kovel, Joel. 2007. *The Enemy of Nature: The End of Capitalism or the End of the World?* (2nd edition). London: Zed Books.

Kumar, Ajay. 2011. Understanding the emerging role of motorcycles in African cities: A political economy perspective. World Bank, Sub-Saharan Africa Transport Policy Program discussion paper No. 13, April.

Kunstler, James Howard. 1993. *The Geography of Nowhere: The Rise and Decline of America's Man-Made Landscape.* New York: Touchstone Book.

Ladd, Brian. 2008. *Autophobia: Love and Hate in the Automotive Age.* Chicago: University of Chicago Press.

Lah, Oliver, Alexander Sohr, Xiaoxu Bei, Kain Glensor, Hanna Hueing, and Miriam Mueller. 2014. Metasys, sustainable mobility for megacities—traffic management and low-carbon transport for Hefei, China. In *Mobility and Transportation: Concepts for Sustainable Transportation in the Future.* Wulf-Holger Arndt, ed, Pp. 94–106. Berlin: Jovis.

Laird, Philip and Peter Newman. 2001. How we got here: The role of transport in the development of Australia and New Zealand. In *Back on Track: Making Transport Policy in Australia and New Zealand.* Philip Laird and Peter Newman, eds. Pp. 1–21. Sydney: UNSW Press.

Lane, Bradley. 2013. Public transportation as a solution to oil dependence. In *Transport Beyond Oil: Policy Choices for a Multimodal Future.* John L. Renne and Billy Fields, eds. Pp. 107–125. Washington, DC: Island Press.

Lane, Clayton, Heshuang Zeng, Chhavi Dhingra, and Aileen Carrigan. 2015. *Carsharing: A Vehicle for Sustainable Mobility in Emerging Markets.* World Resources Institute Center for Sustainable Cities, wricities.org.

Lansbury, Russell D., Jacob Saulwick, and Chris F. Wright. 2008. Globalization and employment relations in the Australian automotive industry. In *Globalization and Employment Relations in the Auto Assembly Industry: A Study of Seven Countries.* Roger Blanpain, ed. Pp. 13–34. Austin, TX: Wolters Kluwer.

Lanzendorf, Martin and Annika Busch-Geertsema. 2014. The cycling boom in large German cities—empirical evidence for successful cycling campaigns. *Transport Policy* 36: 26–33.

Larsen, Jonas, Kay W. Axhausen, and John Urry. 2006. Geographies of social networks: Meetings, travel and communications. *Mobilities* 1:261–283.

Lauier, Eric. 2005. Doing office work on the motorway. In *Automobilities*. Mike Featherstone, Nigel Thrift, and John Urry, eds. Pp. 261–277. London: Sage.

Laurier, Eric et al. 2008. Driving and "passengering": Notes on the ordinary organization of car travel. *Mobilities* 3:1–23.

Law, Christopher. 1991. Motor vehicle manufacturing: The representative industry. In *Restructuring the Global Automobile Industry: National and Regional Impacts.* Christopher M. Law, ed. Pp. 1–18. London: Routledge.

Lee, Robert. 2010. *Transport: An Australian History.* Sydney: UNSW Press.

Legacy, Crystal, David Ashmore, Jan Scheurer, John Stone, and Carey Curtis. 2018. Planning the driverless city. *Transport Reviews* 39:84–102.

Leggett, Jeremy. 2005. *Half Gone: Oil, Gas, Hot Air and the Global Energy Crisis.* London: Portobello Books.

Leung, G.C.K. 2010. China's oil use, 1990-2008. *Energy Policy* 38:932–944.

Lewis, Chris. 2010. Will logistics help end a long winter in Eastern Europe? *Automotive Logistics*, April 1, pp. 46–49.

Lewis, David L. 1983. Sex and the automobile: From rumble seats to rockin' vans. In *The Automobile and American Culture.* David L. Lewis, eds. Pp. 123–136. Ann Arbor: University of Michigan Press.

Lewis, Tom. 1997. *Divided Highways: Building the Interstate Highways, Transforming American Life.* New York: Viking.

Li, Minqi. 2016. *China and the 21st Century.* London: Pluto Press.

Li, Tiebei, Jago Dodson, and Neil Sipe. 2015. Differentiating metropolitan transport disadvantage by mode: Household expenditure on private vehicle fuel and public transport in Brisbane, Australia. *Journal of Transport Geography* 49:16–25.

Ling, Peter J. 1990. *America and the Automobile: Technology, Reform and Social Change.* Manchester: Manchester University Press.

Lloyd, Mike. 2017. On the way to cycle rage: Disputed mobile formations. Mobilities 12: 384–404.

Loewy, Michael. 2015. *Ecosocialism: A Radical Alternative to Capitalist Catastrophe.* Chicago: Haymarket.

Loffler, Don. 2012. *Holden Days.* Kent Town, South Australia: Wakefield Press.

Low, Nicholas. 2013. From automobility to sustainable transport. In *Transforming Urban Transport: The Ethics, Politics and Practices of Sustainable Mobility.* Nicholas Low, ed. Pp. 57–84. London: Earthscan.

Low, Nicholas, Brendan Gleeson, Ray Green, and Darko Radovic. 2005. *The Green City: Sustainable Homes, Sustainable Suburbs.* Sydney: University of New South Wales Press.

Low, Nicholas and Kevin O'Connor. 2013. The dilemma of mobility. *Transforming Urban Transport: The Ethics, Politics and Practices of Sustainable Mobility.* Low, Nicholas, ed. Pp. 3–25. London: Earthscan.

Luce, Mathias. 2015. Sub-imperialism, the highest stage of dependent capitalism. In *BRICS: An Anti-Capitalist Critique.* Patrick Bond and Ana Garcia, eds. Pp. 27–44. London: Pluto Press.

Luethje, Boy. 2014. Labour relations, production regimes and labour conflicts in the Chinese automotive industry. *International Labour Review* 153:535–560.

Luger, Stan. 2000. *Corporate Power, American Democracy, and the Automobile Industry.* Cambridge, UK: Cambridge University Press.

Lutin, Jerome M. 2018. Not if, but when: Autonomous driving and the future of transit. *Journal of Public Transportation.* 21: 92–103.

Lutz, Catherine. 2014. The U.S. car colossus and the production of inequality. *American Ethnologist* 41:232–245.

Lynch, J. and G. Davey Smith. 2005. A life course approach to chronic disease epidemiology. *Annual Review of Public Health* 26:1–35.

Lynd, Robert S. and Helen Merrell Lynd. 1929. *Middletown: A Study in American Culture.* New York: Harvest Books.

———. 1937. *Middletown in Transition: A Study in Cultural Conflict.* New York: Harcourt, Brace.

Magdoff, Fred. 2008. The political economy and ecology of biofuels. *Monthly Review.* July-August, Pp. 34–50.

Magdoff, Fred and John Bellamy Foster. 2010. What every environmentalist needs know about capitalism. *Monthly Review* 61(10):1–30.

———. 2011. *What Every Environmentalist Needs to Know About the Environment.* New York: Monthly Review Press.

Mage, David and Olivier Zali. 1992. Executive summary. In *Motor Vehicle Air Pollution: Public Health Impact and Control Measures.* David Mage and Olivier, eds. Pp. vii-ix. Geneva: World Health Organization and Geneva Department of Public Health.

Maizlish, Neil et al. 2013. Health cobenefits and transportation-related reductions in greenhouse gas emissions in the San Francisco Bay Area. *American Journal of Public Health*, published online ahead of print, February 14.

Malm, Andreas. 2016. Fossil Capital: The Rise of Steam Power and the Roots of Global Warming. London: Verso.

MarketLine 2016. New Cars in South Africa. MarketLine Industry Profile, August. www.marketline.com.

Marquet, Oriol and Carme Miraelles-Guasch. 2016. City of motorcycles: On how objective and subjective factors are behind the rise of two-wheeled mobility in Barcelona. *Transport Policy* 52: 37–45.

Marsh, Peter and Peter Collett. 1986. *Driving Passion: The Psychology of the Car.* London: Jonathan Cape.

Martin, George. 2002. Grounding social ecology: Landscape, settlement, and right of way. *Capitalism Nature Socialism* 13(1):3–30.

Martin, Elliot W. and Susan A. Shaheen. 2010. Greenhouse gas emission impacts of carsharing in North America. June. San Jose, CA: Mineta Transportation Institute, San Jose University.

Matthews, Graham. 2009. Cities Drowning in Cars: What's the Alternative? *Green Left Weekly*, Issue 800, June 29, https://www.greenleft.org.au/content/cities-drowning-cars-whats-alternative.

McCarthy, Tom. 2007. *Auto Mania: Cars, Consumers, and the Environment.* New Haven, CT: Yale University Press.

McLaughlin, Andrew M. and William A. Maloney. 1999. *The European Automobile Industry: Multi-Governance, Policy and Politics.* London: Routledge.

McLuhan, Marshall. 1964. *The Mechanical Bride: Folklore of Industrial Man.* Boston: Beacon Press.

McManus, M.C. 2012. Environmental consequences of the use of batteries in low carbon systems: The impact of battery production. *Applied Energy* 93: 288–295.

McNeill, J.R. and Peter Engelke. 2014. *The Great Acceleration: An Environmental History of the Anthropocene since 1945.* Cambridge, MA: Belknap Press of Harvard University Press.

McGuirk, Justin. 2014. *Radical Cities: Across Latin America in Search for New Architecture.* London: Verso.

Mees, Paul. 2010. *Transport for Suburbia: Beyond the Automobile.* London: Earthscan.

Mees, Paul and Lucy Groehart. 2014. Canberra's forgotten transport history. In *The Public City: Essays in Honour of Paul Mees*, eds. Pp. 182–194. Melbourne: Melbourne University Press.

Mehndiratta, S. and A. Salzberg. 2012. Improving public transport in Chinese cities: Elements of an action plan. In *Sustainable Low-Carbon City Development in China.* A. Baeumler, E. Ijjasz-Vasquez, and S. Mehndirrata, eds. Pp. 269–298. Washington, DC: World Bank.

Mercier, Jean, Mario Carrier, Fabio Duarte, and Fanny Tremblay-Racicot. 2016. Policy tools for sustainable transport in three cities in the Americas: Seattle, Montreal and Curitiba. *Transport Policy* 50: 95–105.

Metz, David. 2008. *The Limits to Travel: How Far Will You Go?* London: Earthscan.

———. 2010. *Controlling Climate Change.* Cambridge, UK: Cambridge University Press.

———. 2013. Peak car and beyond: The fourth era of travel. *Transport Reviews* 33:255–270.

Meyer, William B. 2013. *The Environmental Advantages of Cities: Countering Commonsense Antiurbanism.* Cambridge, MA: MIT Press.

Michael, Mike. 2001. The invisible car: The cultural purification of road rage. In *Car Cultures.* Daniel Miller, eds. Pp. 59–80. Oxford, UK: Berg.

Mikler, John. 2009. *Greening the Car Industry: Varieties of Capitalism and Climate Change.* Cheltenham, UK: Edward Elgar.

Miliband, Ralph. 1994. *Socialism for a Skeptical Age.* London: Courage Press.

Millard-Ball, A et al. 2005. Car-sharing: Where and how it succeeds. Transit Cooperative Research Program Report 108. Washington, DC: Federal Transit Administration.

Miller, Daniel. 2001. Driven societies. In *Car Cultures.* Daniel Miller, ed. Pp. 1–33. Oxford, UK: Berg.

Miller, J. Hillis. 2006. Virtual automobility: Two ways to get a life. *Sociological Review*, Supplement 1: 193–207.

Mills, Kevin. 2013. Health, oil-free transportation: The role of walking and bicycling in reducing oil dependence. In *Transport Beyond Oil: Policy Choices for a Multimodal Future.* John L. Renne and Billy Fields, ed. Pp. 178–187. Washington, DC: Island Press.

Minchin, Timothy. 2007. "The assembly line and cars come first": Labor relations and the demise of the Nissan car manufacturing in Australia. *Labor History* 48:327–346.

Moeser, Kurt. 2001. *Autobasteln*; Modifying, maintaining, and repairing private cars in the GDR, 1970–1990. In *The Socialist Car: Automobility in the Eastern Bloc.* Lewis H. Siegelbaum, ed. Pp. 157–169. Ithaca, NY: Cornell University Press.

———. 2003. The dark side of "automobilism," 1900–1930: Violence, war and the motor car. *Journal of Transport History* 24(2):238–258.

Mokhtarian, Partricia L., Ilan Salomon, and Matan E. Singer. 2015. What moves us? An interdisciplinary exploration for traveling. *Transport Reviews* 35 (3): 251–274.

Mom, Gijs. 2015. *Atlantic Automobilism: Emergence and Persistence of the Car 1985–1940.* New York: Berghahn.

Moorhouse, H.F. 1983. American Automobiles and Workers' Dreams. *Sociological Review* 31:403–426.

Montgomery, Scott L. 2010. *The Powers That Be: Global Energy for the Twenty-First Century and Beyond.* Chicago: University of Chicago Press.

Moraabia, Alfredo et al. 2010. Potential health impact of switching from car to public transportation when commuting to work. *American Journal of Public Health* 100 (12): 2388–2391).

Morales, Rebecca. 1994. *Flexible Production: Restructuring of the International Automobile Industry.* London: Polity Press.

Morgan, Faith, Eugene Murphy, Megan Quinn, and Bruce Cromer. 2006. *The Power of Community: How Cuba Survived Peak Oil.* Yellow Springs, OH: Community Service.

Moriarty, Patrick and Clive Beed. 1992. The car in its second century. *Journal of Australian Political Economy.* Number 29, pp. 26–39.

Mosey, Chris. 2000. *Car Wars: Battles on the Road to Nowhere.* London: Vision Paperbacks.

Mumford, Lewis. 1963. *The Highway and the City.* New York: Harcourt, Brace & World.

———. 1964. *Understanding Media: The Extensions of Man.* London: Ark Paperbacks.

Mugyenyi, Bianca and Yves Engler. 2011. *Stop Signs: Cars and Capitalism.* Vancouver, BC: RED Publishing.

Mumford, Lewis. 1964. *The Highway and the City* (Revised edition). London: Secker & Warburg.

Myer, John M. 2015. *Engaging the Everyday: Environmental Social Criticism and the Resonance Dilemma.* Cambridge, MA: MIT Press.

Nader, Ralph. 1965. *Unsafe at Any Speed: The Designed-In Dangers of the American Automobile.* New York: Greenwood Press.

Naegele, Jolyon. 2004. Ten years after—the East German Trabant. In *Ten Years After: The Fall of Communism in East/Central Europe.* http://interconnected.org/notes/2004/11/prague/mirrors/www.rferl.org/nca/special/10years/germany3.html

Naess, Petter. 2006. *Urban Structure Matters: Residential Location, Car Dependence and Travel Behaviour.* London: Routledge.

Narotzky, Viviana. 2002. Our cars in Havana. In *Autopia: Cars and Culture.* Peter Wollen and Joe Kerr, eds. Pp. 169–176. London: Reaktion Books.

Neale, Jonathan. 2008. *Stop Global Warming: Change the World.* UK: Bookmark Publications.

Next System Project. 2016. A conversation with Noam Chomsky on organizing for a next system. http://thenextsystem.org/conversation_with_noam_chomsky/

Newman, Peter. 2009. Transport opportunities: Towards a resilient city. In *Opportunities Beyond Carbon: Looking for a Sustainable World.* John O'Brien, ed. Pp. 98–115. Melbourne: Melbourne University Press.

Newman, Peter and Christy Newman. 2012. Car. In *Sociology: Antipodean Perspectives.* Peter Beilharz and Treveor Hogan, eds. Pp. 358–363. Sydney: Oxford University Press.

Newman, Peter and Jeffrey Kenworthy. 1999. *Sustainability and Cities: Overcoming Automobile Dependency.* Washington, DC: Island Press.

———. 2000. The ten myths of automobile dependence. *World Transport Policy & Practice* 6(1):15–25.

———. 2006. Urban design to reduce automobile dependence. *Opolis* 2(1):35–52.

———. 2015. *The End of Automobile Dependence: How Cities are Moving Beyond Car-Based Planning.* Washington, DC: Island Press.

Newman, Peter, Jeffrey Kenworthy, and Mark Bachels. 2001. How we compare: Patterns and trends in Australian and New Zealand cities. In *Rethinking Transport Policy in Australia and New Zealand.* Philip Laird, Peter Newman, Mark Machels, and Jeffrey Kenworthy, eds. Pp. 45–67. Sydney: UNSW Press.

Newman, Peter, Jeff Kenworthy, and Garry Glazebrook. 2013. Peak car use and the rise of global rail: Why this is happening and what it means for large and small cities. *Journal of Transportation Technologies* 3:272–287.

Newman, Peter and Isabella Jennings. 2008. *Cities as Sustainable Ecosystems: Principles and Practices.* Washington, DC: Island Press.

Newman, Peter, Timothy Beatley, and Heather M. Boyer. 2017. *Resilient Cities: Overcoming Fossil-Fuel Dependence.* Gabriola, BC: Island Press.

Newton, P. J., S. Plotkin, D. Sperling, R. Wit, and P. J. Zhou, 2007: Transport and its infrastructure. In Climate Change 2007. Mitigation. Contribution of Working Group III to the Fourth Assessment Report of the Intergovernmental Panel on Climate Change. B. Metz, O.R.

Next System Project. 2016. A Conversation with Noam Chomsky on Organizing for a Next System. March 24, https://thenextsystem.org/conversation_with_noam_chomsky.

Nieuwenhuis, Paul. 2014. *Sustainable Aotomobility.* Rockville, MD: Edgaronline.

Nielsen, Kenneth and Harold Withite. 2015. The rise and fall of the "people's car": middle-class aspirations, status and mobile symbolism in "*New India.*" *Contemporary South Asia* 23:371–387.

Norton, Peter. 2008. *The Dawn of the Motor Age in the American City.* Oxford, UK: Oxford University Press.

Notar, Beth. 2017. Car crazy: The rise of car culture in China. In *Cars, Automobility and Development in Asia: Wheels of Change.* Arve Hansen and Kenneth Bo Nielsen, eds. Pp. 152–170. London: Routledge.

O'Connell, J., and A. Myers. 1966. *Safety Last: An Indictment of the Auto Industry.* New York: Random House.

O'Connell, Sean. 1998. *The Car in British Society: Class, Gender, and Motoring, 1896–1939.* Manchester: Manchester University Press.

O'Dell, Tom. 2001. *Raggare* and the panic of mobility: Modernity and everyday life in Sweden. In *Car Cultures*. Daniel Miller, eds. Pp. 105–132. Oxford, UK: Berg.

Odd spot. The Age (Melbourne), April 17, 2017, p. 1.

Opinion Research Centre. 1951. Survey of movement within Great Melbourne. Unpublished report. Melbourne: Melbourne Research Centre.

Oppe, Siem. 1991. The development of traffic and traffic safety in six developed societies. *Accident Analysis and Prevention* 23:401–412.

Orssatto, R. and S. Clegg. 1999. The political ecology of organisations. *Organization and Environment* 12:263–279.

Owen, David. 2009. *Green Metropolis: Why Living Smaller, Living Closer, and Driving Less Are the Keys to Sustainability*. New York: Riverhead Books.

Packer, Jeremy. 2008. *Mobility without Mayhem: Safety, Cars, and Citizenship*. Durham, NC: Duke University Press.

Parsons, Howard L., ed. 1977. *Marx and Engels on Ecology*. Westport, CT: Greenwood Press.

Parissien, Steven. 2013. *The Life of the Automobile: A New History of the Motor Car*. London: Atlantic Books.

Partnership for Sustainable Urban Transport in Asia. n.d. Sustainable urban transport in Asia: Making the vision a reality. A CAI-Asia Program. http://pdf.wri.org/sustainable_urban_transport_asia.pdf.

Paskal, Cleo. 2010. *Global Warring: How Environment, Economic, and Political Crises Will Redraw the World Map*. New York: Palgrave Macmillan.

Paterson, Matthew. 2007. *Automobile Politics: Ecology and Cultural Political Economy*. Cambridge, UK: Cambridge University Press.

———. 2014. Governing mobilities, mobilising carbon. *Mobilities* 9:570–584.

Patton, Phil. 2002. *Bug: The Strange Mutation of the World's Most Famous Automobile*. New York: Simon & Schuster.

Pedersen, Morten Axel and Mikkel Bunkenborg. 2012. Roads that separate: Sino-Mongolian relations in the inner Asian desert. *Mobilities* 7:555–569.

Pendakur, V Setty. 2011. Non-motorized urban transport in neglected modes. In *Urban Transport in the Developing World: A Handbook of Policy and Practice*. Harry T. Dimitrious and Ralph Gakenheimer, eds. Pp. 203–261. Cheltenham, UK; Edward Elgar.

Penna, Anthony N. 2015. *The Human Footprint: A Global Environmental History*. Malden, MA: Wiley Blackwell.

Peters, Peter Frank. 2006. *Time, Innovation and Mobilities: Travel in Technological Cultures*. London: Routledge.

Petersen, Tim. 2016. Watching the Swiss: A network approach to rural and exurban public transport. *Transport Policy* 52:175–185.

Piemtel, David. 2010. Biofuels versus food resources and the environment. *Review* 33 (2/3): 177–201.

Pietri F., A. Leclerc, L. Boitel, J.F. Chastang, J.F. Morcet J.F., and M. Blondet. 1992. Low-back Paain in Commerical Travelers. Scandinavian *Journal of Work and Environmental Health* 18:52–58.

Pinch, Philip and Suzanne Reimer. 2012. Moto-mobilities: Geographies of the motorcycle and motorcyclists. *Mobilities* 7:439–457.

Pirani, Simon. 2018. *Burning Up: A Global History of Fossil Fuel Consumption.* London: Pluto.

Pointing, Clive. 1992. *A Green History of the World: The Environment and the Collapse of Great Civilizations.* New York: St. Martin's Press.

Pooley, Colin G. et al. 2012. The role of walking and cycling in reducing the impacts of climate change. In *Transport and Climate Change.* Tim Ryley and Lee Chapman, eds. Pp. 175–195. Bingley, UK: Emerald Group.

Porter, Gina. 2014. Transport services and their impact on poverty and growth in rural sub-Saharan Africa: A review of recent research and future research needs. *Transport Reviews* 34:25–45.

Porter, Richard C. 1999. *Economics at the Wheel: The Cost and Drivers.* San Diego, CA: Academic Press.

Prud'homme, Remy and Martin Koning. 2012. Electric vehicles: A tentative economic and environmental evaluation. *Transport Policy* 23: 60–69.

Psarafitis, H.N. and C.A. Kontovas. 2009. CO_2 emission statistics for the world commercial fleet. *WMU Journal of Maritime Affairs* 8:1–25.

Quellet, Lawrence J. 1994. *Pedal to the Metal: The Work Lives of Truckers.* Philadelphia: Temple University Press.

Quinn, Jr., Dennis Patrick. 1988. *Restructuring the Automobile Industry: A Study of Firms and States in Modern Capitalism.* New York: Columbia University Press.

Rahul, T.M. and Ashish Verma. 2014. A study of acceptable trip distances using walking and cycling in Bangalore. *Journal of Transport Geography* 38:106–113.

Rajan, Sudhir Chella. 2006. Automobility and the liberal disposition. *Sociological Review*, Supplement 1:113–129.

Raji, Rafiq. 2018. African Auto Sector Motors Ahead. *African Business*, August/September: 52–54.

Randall, Richard. 2017. The microsociology of automobility: The production of the automobile self. *Mobilities* 12:663–676.

Ray, Manas Ranjan and Twisha Lahiri. 2010. Health effects of urban air pollution in India. In *Air Pollution: Health and Environmental Impacts.* Bhola R. Gurjar, Luisa T. Molina, and Chandra S.P. Ojhas, eds. Pp. 165–201. Boca Raton, FL: CRC Press.

Reddy, B. Sudhakara and P. Balachandra. 2012. Urban mobility: A comparative analysis of megacities of India. *Transport Policy* 21: 152–164.

Register, Richard. 2001. *Ecocities: Building Cities in Balance with Nature.* Berkeley, CA: Berkeley Hills Books.

———. 2011. Green cities must serve people instead of cars. In *Green Cities: At Issue Environment.* Ronald D. Lankford, Jr., ed. Pp. 73–80. Detroit: Gale. Glenage Learning.

Reichert, Alexander, Christian Holz-Rau, and Joachim Scheiner. 2016. GHG emissions in daily travel and long-distance travel in Germany—social and spatial correlates. *Transport Research* Part D 49:25–43.

Renner, Michael. 2008. Vehicle production rises sharply. In *Vital Signs 2007–2008: Trends That Are Shaping Our Future.* Worldwatch Institute, ed. Pp. 66–67. New York: W.W. Norton & Company.

Redshaw, Sarah. 2007. Articulations of the car: the dominant articulations of racing and rally driving. *Mobilities* 2:121–141.

Reese, Katherine G. 2016. Accelerate, reverse, or find the off ramp? Future automobility in the fragmented American imagination. *Mobilities* 11:152–170.

Richter, Burton. 2010. *Beyond Smoke and Mirrors: Climate Change and Energy in the 21st Century*. Cambridge, UK: Cambridge University Press.

Riley, Robert Q. 1994. *Alternative Cars in the 21st Century: A New Personal Transportation Paradigm*. Warrendale, PA: Society of Automotive Engineers.

Rogers, Richard. 1997. *Cities for a Small Planet*. London: Faber and Faber.

Ross, H. Laurence. 1992. *Confronting Drunk Driving: Social Policy for Saving Lives*. New Haven, CT: Yale University Press.

Robbins, Paul, John Hintz, and Sarah A. Moore. 2010. *Environment and Society*. Malden, MA: Wiley-Blackwell.

Rosenfeld, Herman. 2009. The North American auto industry in crisis. *Monthly Review*, June, pp. 18–36.

Rothe, J. Peter. 2008. *Driven to Kill: Vehicles as Weapons*. Edmonton: University of Alberta Press.

Rosenfeld, Herman. 2009. The North American auto industry in crisis. *Monthly Review*, June, pp. 18–36.

Rubin, Eli. 2011. Understanding a car in the context of a system: Trabants, Marzahn, and East German socialism. In *The Socialist Car: Automobility in the Eastern Bloc*. Lewis H. Siegelbaum, ed. Pp. 124–140. Ithaca, NY: Cornell University Press.

Rutledge, Ian, 2006. *Addicted to Oil: America's Relentless Drive for Energy Security*. London: I.B. Tauris.

Rutherford, Tod D. and John Holmes. 2014. Manufacturing resiliency: Economic restructuring and automotive manufacturing in the Great Lakes region. *Cambridge Journal of Regions, Economy and Society* 7:359–378.

Rylander, Ragnar. 1992. Effects on humans of environmental noise particularly from road traffic. In *Motor Vehicle Pollution: Public Health Impact and Control Measures*. David Mage and Olivier Zali, eds. Pp. 63–83. Geneva: World Health Organization and Geneva Department of Public Health.

Sachs, Jeffrey. 2008. *Common Wealth: Economics for a Crowded Planet*. New York: Allen Lane.

Sachs, Wolfgang (translated by Don Reneau). 1992. *For Love of the Automobile: Looking Back into the History of Our Desires*. Berkeley: University of California Press.

Sanders, Barry. 2009. *The Green Zone: The Environmental Costs of Militarism*. Oakland, CA: AK Press.

Sandra. 2010. The "Trabant": The "Volkswagen' of the former German Democratic Republic. Posted on July 29 on Culture, History, Travel.

Sanford, Jim. 2016. Auto Shutdown Will Deliver Another Economic Blow. *Australia Institute Briefing Notes*. October. https://www.tai.org.au/sites/default/files/Auto_Shutdown_Will_Deliver_Another_Economic_Blow.pdf.

Seow, Victor. 2014. Socialist drive: The First Auto Works and the contradictions of connectivity in the early People's Republic of China. *Journal of Transport History* 35(2):145–161.

Sernau, Scott. 2009. *Global Problems: The Search for Equity, Peace, and Sustainability* (2nd edition). Boston: Allyn & Bacon.

Schaefer, Andreas. 2011. The future of energy for urban transport. In *Urban Transport in the Developing World: A Handbook of Policy and Practice*. Harry T. Dimitrious and Ralph Gakenheimer, eds. Pp. 113–136. Cheltenham, UK: Edward Elgar.

Schaefer, Andreas, John B. Heywood, Henry D. Jacoby, and Ian A. Waitz. 2009. *Transportation in a Climate-Constrained World*. Cambridge, MA: MIT Press.

Schaefer, Tanja and Angela Jain. 2014. Energy-efficient transport planning for Hyderabad, India: challenges: Planning in a fast-changing environment. In *Mobility and Transportation: Concepts for Sustainable Transportation in the Future*. Arndt, Wulf-Holger ed. Pp. 39–57. Berlin: Jovis.

Schiller, Preston L., Eric C. Bruun, and Jeffrey R. Kenworthy. 2010. *An Introduction to Sustainable Transportation: Policy, Planning and Implementation*. London: Earthscan.

Schmucki, Barbara. 2003. Cities as traffic machines: Urban transport planning in East and West Germany. In *Suburbanizing the Masses: Public Transport and Urban Development in Historical Perspective*. Colin Dival and Winstan Bond, eds. Pp. 149–170. Aldershot, England: Ashgate.

Schipper, Lee. 2011. Automobile use, fuel economy and CO_2 emissions in industrialized countries: Encouraging trends through 2008. *Transport Policy* 18:358–372.

Schmidt-Relenberg, Norbert. 1986. On the sociology of car traffic in towns. In *Transport Sociology: Social Aspects of Transport Planning*. Enne de Boer, ed. Pp. 121–132. Oxford, UK: Pergamon Press.

Scheinbaum, Claudia. 2009. Environmental solar heating standard: GHG mitigation policy in Mexico City. In Green CIYYnomics: The Urban War against Climate Change. Kenny Tang, ed. Pp. 2006:206-219. Sheffield, UK: Greenleaf Publishing.

Schipper, Lee, Elizabeth Deakin, and Carolyn McAndrews. 2011. Carbon dioxide emissions from road transport in Latin America. In *Transport Moving to Climate Intelligence*. Werner Rothengatter et al., eds. Pp. 111–127. New York: Springer.

Schweid, Richard. 2004. *Che's Chevrolet, Fidel's Oldsmobile: On the Road in Cuba*. Chapel Hill: University of North Carolina Press.

Schwaegerl, Christian. 2014. *The Anthropocene: The Human Era and How It Shapes Our Planet*. Santa Fe: Synergetic Press.

Seiler, Cotton. 2008. *Republic of Drivers: A Cultural History of Automobility in America*. Chicago: University of Chicago Press.

Shahan, Zachary. 2010. NASA says: Automobiles largest net climate change culprit. *Scientific American*, February 24.

Sheller, Mimi. 2014. *Aluminum Dreams: The Making of Light Modernity*. Cambridge, MA: MIT Press.

———. 2018. *Mobility Justice: The Politics of Movement in an Age of Extremes*. London: Verso.

Shiva, Vandana. 2008. *Soil Not Oil: Environmental Justice in an Age of Climate Crisis*. Boston: South End Press.

Short, John Rennie and Luis Mauricio Pinet-Peralta. 2010. No accident: Traffic and pedestrians in the modern city. *Mobilities* 5:41–59.

Shotwell, Gregg. 2015. Manufacturing America's dreams. *Monthly Review* 67(1): 36–40.
Siegelbaum, Lewis H. 2008. *Cars for Comrades: The Life of the Soviet Automobile.* Ithaca, NY: Cornell University Press.
———. 2011. Introduction. In *The Socialist Car: Automobility in the Eastern Block.* Lewis H. Siegelbaum, ed. Pp. 1–13. Ithaca, NY: Cornell University Press.
Sims, Ralph et al. 2014. Transport. In *Climate Change 2014: Mitigation of Climate Change.* Contribution of Working Group III to the Fifth Assessment Report of the Intergovernmental Panel on Climate Change. O. Edenhofer et al., eds. Pp. 599–670. Cambridge, UK: Cambridge University Press.
Singleton, Patrick A. 2019. Discussing the "positive utilities" of autonomous vehicles: Will travellers really use their time productively. *Transport Reviews* 39:5–065.
Skippon, Stephen and Nick Reed. 2017. The future of transport? The psychology of self-driving vehicles. *The Psychologist* 30:22–27.
Slowik, Peter and Nic Lutsey. 2017. Expanding the electric vehicle market in U.S. cities. International Council on Clean Transportation White Paper, July, http://www.theicct.org.
Smeed, R.J. 1949. *Journal of the Royal Statistical Society* A(I): 1–34.
Smith, E.O. 2002. *When Culture and Biology Collide: Why We Are Stressed, Depressed, and Self-Obsessed.* New Brunswick, NJ: Rutgers University Press.
Sousanis, John. 2011. World vehicle population tops 1 billion units. http://wardsauto.com/.../world-vehicle-population-tops-1-billion-units.
Sovacool, B.K. 2010. A Transition to Plug-in Hybrid Electric Vehicles (phevs): Why Public Health Professionals Must Care. *Journal of Epidemiology and Community Health* 64:185–187.
Sperling, Daniel and Deborah Gordon, 2010. *Two Billion Cars: Driving Toward Sustainability.* New York: Oxford University Press.
Staley, Sam and Adrian Moore. 2009. *Mobility First: A New Vision for Transportation in a Globally Competitive Twenty-First Century.* Lanham, MD: Rowman & Littlefield.
Standard, Jim. 2016. Manufacturing (still) matters: Why the decline of Australian manufacturing is NOT inevitable, and what government can do about it. Briefing paper, June. Canberra: The Australia Institute.
Stanford, Jim. 2008. *Economics for Everyone: A Short Guide to the Economics of Capitalism.* London: Pluto Press.
Stilwell, Frank J. B. 1992. *Understanding Cities & Regions: Spatial Political Economy.* Sydney: Pluto Press.
Streeck, Wolfgang. 2016. *How Will Capitalism End? Essays on a Failing System.* London: Verso.
Stokes, Gordon. 2013. The prospects for future levels of car access and use. *Transport Review* 33:360–375.
Stone, John and Beau B. Beza. 2014. Public transport: Elements for success in a car-oriented city. In *The Public City: Essays in Honour of Paul Mees.* Brendan Gleeson and Beau B. Beza, eds. Pp. 90–105. Melbourne: Melbourne University Press.

Stone, John and Paul Mees. 2016. Public transport networks in the post-petroleum era. In *Plarning After Petroleum. Planning After Petroleum: Preparing Cities for the Age Beyond Oil.* Jago Dodson, Neil Sipe, and Anitra Nelson, eds. Pp. 113–128. London: Routledge.

Stotz, Gertrude. 2001. The colonizing vehicle. In Car Cultures. Daniel Miller, ed. Pp. 223–244.

Stover, John F. 1999. *The Routledge Historical Atlas of the American Railroads.* New York: Routledge.

Stroemberg, Helena, I.C. MariAnne Karlsson, and Oskar Rexfelt. 2015. Eco-driving: Drivers' understanding of the concept and implications of future interventions. *Transport Policy* 30: 48–54.

Sutter, Paul S. 2002. *Driven Wild: How the Fight Against Automobiles Launched the Modern Wilderness Movement.* Seattle: University of Washington Press.

Sweezy, Paul. 1973. Cars and Cities. *Monthly Review* 24(11):1–18.

Taebel, Delbert A. and James V. Cornehls. 1977. *The Political Economy of Urban Transportation.* Port Washington, NY: Kennikat Press.

Tetzlaff, Stefan. 2017. Revolution or evolution? The making of the automobile sector as a key industry in mid-twentieth century India. In *Cars, Automobility and Development in Asia: Winds of Change.* Arve Hansen and Kenneth Bo Nielsen, eds. Pp. 62–80. London: Routledge.

Tian, Yihui, Qinghua Zhu, Kee-hung Lai, and Y.H. Venus Lun. 2014. Analysis of greenhouse gas emissions of freight transport sector in China. *Journal of Transport Geography* 40:43–52.

Tickell, Oliver. 2008. *Kyoto2: How to Manage the Global Greenhouse.* London: Zed Books.

Timilsina, Govinda R. 2009. Factors affecting transport CO_2 emissions growth in Latin American and Caribbean countries: An LMDI decomposition analysis. *International Journal of Energy Research* 33:396–414.

Thomas, David. 2018. Ignition Time? Africa's Automotive Sector in Focus. *African Business*, March: 12–19.

Todorovich, Petra and Edward Burgess. 2013. High-speed rail and reducing oil dependence. In *Transport Beyond Oil: Policy Choices for a Multimodal Future.* John L. Renne and Billy Fields, ed. Pp. 141–160. Washington, DC: Island Press.

Tomaschek, Jan. and Ulrich Fahl. 2014. Climate-protection strategies for the transport sector of Gauteng Province, South Africa. In *Mobility and Transportation: Concepts for Sustainable Transportation in the Future.* Arndt, Wulf-Holger ed. Pp. 78–92. Berlin: Jovis.

Tomba, Luigi. 2014. *The Government Next Door: Neighborhood Politics in Urban China.* Ithaca, NY: Cornell University Press.

Trainer, Ted. 1998. *Saving the Environment: What Will It Take.* Sydney: UNSW Press.

———. 2010. *The Transition to a Sustainable and Just World.* Canterbury, New South Wales: Environbook.

———. 2015. Remaking settlements: The potential cost reductions enabled by the simpler way. Simplicity Institute Report 15e.

Tribbia, John. 2007. Stuck in the slow lane of behaviour change? A not-so-superhuman perspective on getting out of cars. In *Creating a Climate for Change: Communicating Climate and Facilitating Social Change*. Susan C. Moser and Lisa Dilling, eds. Pp. 238–249.

Troy, Patrick. 2010. Urban public transport: A study in commonwealth-state relations. In *The Public City: Essays in Honour of Paul Mees*, eds. Pp. 168–181. Melbourne: Melbourne University Press.

Tziovaras, Theoharis. 2011. Is there such a thing as an environmentally-friendly car? *World Transport Policy and Practice* 17(3):27–31.

Umweltstatistik Schweiz. 2009. Swiss environmental statistics. Swiss Federal Office for the Environment.

United Nations. 2016. Habitat IIII issue papers, 19—transport and mobility. May 31.

US Bureau of Transportation Statistics. 2009. Latest Statistics. https://www.bts.gov/.

US Census Bureau. 2000-2015. *American Fact Finder* (Volume 2010). Washington, DC: Department of Commerce.

US Department of Transportation. 2008. Travel to school: The distance factor. National Household Travel Survey Brief, January, http://nhts,ornl.gov/publications.shtml.

United States Environmental Protection Agency. 2014. Greenhouse gas emissions for a typical passenger vehicle. Washington, DC: Office of Transportation and Air Quality (EPA-420-F-14-040a, May).

United States Environmental Protection Agency. 2015a. Fast facts: U.S. transportation sector greenhouse gas emissions 1990–2012. Washington, DC: Office of Transportation and Air Quality (EPA-420-F-15-0002).

———. 2015b. Fast facts: U.S. transportation sector greenhouse gas emissions 1990–2013. Washington, DC: Office of Transportation and Air Quality (EPA-420-F-15-032), October.

United States Federal Transit Administration. 2009. Public transportation's role in responding to climate change. Washington, DC: U.S. Department of Transportation.

Urry, John. 2000. *Sociology beyond Societies: Mobilities for the Twenty-First Century*. London: Routledge.

———. 2004. The "system" of automobility. *Theory, Culture & Society* 21(4/5):25–39.

———. 2007. *Mobilities*. London: Polity.

———. 2008.—p. 41

———. 2016. Complex systems and multiple crises of energy. In *Anthropology and Climate Change: From Actions to Transformations*. Susan A. Crate and Mark Nuttall, eds. Pp. 105–120. New York: Routledge.

Vahrenkamp, Richard. Automobile tourism and Nazi propaganda: Constructing the Munich-Salzburg autobahn, 1933–1939. *Journal of Transport History* 27(2):21–38.

Van Themsche, S. 2016. *The Advent of Unmanned Electric Vehicles*. Springer International Publishing Switzerland.

Vanderbilt, Tom. 2008. *Traffic: Why We Drive the Way We Do* (and What It Says about Us). London: Penguin.

Vanderheiden, Steve. 2006. Assessing the case against the SUV. *Environmental Politics* 15: 23–40.

Vasconcellos, Eduardo A. 2001. *Urban Transport, Environment and Equity: The Case for Developing Countries.* London: Earthscan.

Vaze, Prashant. 2009. *The Economical Environmentalist: My Attempt to Live a Low-Carbon Life and What It Cost.* London: Earthscan.

Verma, Ashishi and T.V. Ramanayya. 2015. *Public Transport Planning and Managing in Developing Countries.* Boca Raton, FL: CRC Press.

Vilimek, Tomas and Valentina Fava. 2017. The Czechoslovak automotive industry and the launch of a new model: The Skoda factory in Mlada Boleslav, in the 1970s and 1980s. *Journal of Transport History* 38(1): 53–69.

Virilio, Paul. 1991. *The Aesthetics of Disappearance*, translated by Philip Beitchman, New York: Semiotext(e).

———. 1986. *Speed & Politics.* New York: Semiotext(e).

Voelcker, John. 2014. 1.2 billion vehicles on world's roads now, 2 billion by 2035; report. July 29. https://www.greencarreports.com/news/1093560_1-2-billion-vehicles-on-worlds-roads-now-2-billion-by-2035-report, accessed December 27, 2017.

Volti. Rudi. 1996. A century of automobility. *Techology and Culture* 37:663–685.

———. 2004. *Cars and Culture: The Life Story of a Technology.* Westport, CT: Greenwood Press.

———. 2010. For the love of cars. *Journal of Transport History* 28(2):294.

Vovanco, Luis A. 2013. The mundane bicycle and the environmental virtues of sustainable urban mobility. In *Environmental Anthropology: Future Directions.* Helen Kopnina and Eleanore Shoreman-Ouimet, eds. Pp. 25–45. London: Routledge.

Wailes, Nick, Russell D. Lansbury, Jim Kitay, and Anja Kirsch. Globalization, varieties of capitalism and employment relations in the automotive assembly industry. In *Globalization and Employment Relations in the Auto Assembly Industry: A Study of Seven Countries.* Roger Blanpain, eds. (2008). Pp. 1–11. Austin, TX: Wolters Kluwer.

Waitt, Gordon, Theresa Harada, and Michelle Duffy. 2017. 'Let's have some music': Sound, gender and car mobility. *Mobilities* 12: 324–342.

Walsh, Margaret. 2002. Gendering transport history: Retrospect and prospect. *Journal of Transport History* 23(1):1–8.

———. 2008. Gendering mobility: Women, work and automobility in the United States. *Journal of the Historical Association* 93(311): 376–395.

———. 2012. Gender on the road in the United States: By motor car or motor coach? *Journal of Transport History* 31(2):210–230.

Wang, Shu-Mei and Koustuv Dalal. 2012. Road traffic injuries in Shanghai, China. *HealthMED* 6(1):74–80.

Wells, Christopher W. 2013. *Car Country: An Environmental History.* Vancouver, BC: University of Washington Press.

Welzer, Harald. 2012. *Climate Wars: Why People Will Be Killed in the Twenty-First Century* (translated by Patrick Camiller). London: Polity.

Widmer, E.L. 2002. Crossroads: The automobile, rock and roll and democracy. In *Autotopia: Cars and Culture.* Peter Wollen and Joe Kerr, eds. Pp. 65–74. London: Reaktion Books.

Wik, Reynold M. 1972. *Henry Ford and Grass-Roots America*. Ann Arbor: University of Michigan Press.

Wikstroem, Martina, Lisa Hansson, and Per Alvfors. 2016. Investigating barriers for plug-in electric vehicle deployment in fleets. *Transportation Research* Part D 49:59–67.

Williams, Chris. 2010. *Ecology and Socialism: Solutions to Capitalist Ecological Crisis*. Chicago: Haymarket.

Wolf, Winfried. 1996. *Car Mania: A Critical History of Transport*, trans. Gus Fagan. London: Pluto Press.

Wolfe, Joel. 2010. *Autos and Progress: The Brazilian Search for Modernity*. Oxford, UK: Oxford University Press.

Woodcock, James and Rachel Aldred. 2008. Cars, corporations and commodities: Consequences for the social determinants of health. *Emerging Themes in Epidemiology* 5(4): 1–11.

World Health Organization. 2015. *Global Status Report on Road Safety*. Geneva: WHO.

WorldLifeExpectancy. n.d. http://www.worldlifeexpectancy.com/cause-of-death/road traffic- accidents/by-country/.

Worstall, Tim. 2012. The story of Henry Ford's $5 a day wages: It's not what you think. Forbes, March 4, http://www.forbes.com/sites/timworstall/2012/03/04/the-story-of-henry-fords-5-a-day-wages-its-not-what-you-think/.

Wright, Charles L. 1992. *Fast Wheels: Slow Traffic: Urban Transport Choices*. Philadelphia: Temple University Press.

Wright, Erik Olin. 2010. *Envisioning Real Utopias*. London: Verso.

Wright, John. 1998. *Heart of the Lion: The 50 Year History of Australia's Holden*. St. Leonard's NSW: Allen & Unwin.

Xia, Ting, Ying Zhange, Shona Crabb, and Pushan Shah. 2013. Cobenefits of replacing car trips with alternative transportation: a review of evidence and methodological issues. *Journal of Environmental and Public Health* 2013: 797312. Published online 2013 July 16. doi: 10.1155/2013/797312.

Yacobucci, Brent D., Bill Canis, and Richard L. Lattanzio. 2012. Automobile and truck fuel economy (CAFÉ) and greenhouse gas standards. Washington, DC: Congressional Research Service, September 11.

Yago, Glenn. 1984. *The Decline of Transit: Urban Transportation in German and U.S. Cities, 1900–1970*. Cambridge, UK: Cambridge University Press.

Yates, Brock. 1999. *Outlaw Machine: Harley-Davidson and the Search for the American Soul*. Boston: Little, Brown and Company.

Yedla, Sudhakar, Ram M. Shrestha, and Gabriel Anandarajah. 2005. Environmentally sustainable urban transportation—comparative analysis of local emission mitigation strategies vis-à-vis GHG mitigation strategies. *Transport Policy* 12: 245–254.

Young, Diana. 2001. The life and death of cars: Private vehicles on the Pitjantjatjara lands, South Australia. In *Car Cultures*. Daniel Miller, ed. Pp. 35–57. Oxford, UK: Berg.

Zuckermann, Wolfgang. 1991. *The End of the Road: From World Car Crisis to Sustainable Transportation*. Post Mills, VT: Chelsea Publishing Company.

Index

accidents, motor vehicle, 1, 52, 61, 63, 113–20, 122, 124–34, 178, 180, 199, 207
aerotropolis, 107
African Americans, 38, 52, 57, 70, 116, 170
air pollution, 9, 80, 82, 86, 89–95, 98–101, 104, 199, 131, 161, 170, 175–78, 183
air ships, 1, 9
America Walks, 187
American Automobile Association (AAA), 51, 70
American Lung Association, 92
anti-freeway movement, 153, 184
anti-SUV movement, 182
anti-systemic movement, 194
Association of American Railroads, 36, 69, 105
auto formation, 13
automobile or car dependency, 9, 70–71, 74, 148–49, 153, 171
automobile cities, 70, 78
automobile shows, 55
automobilization, 6, 10, 17, 33, 133, 164, 185, 211
autonomous vehicles (AVs), 179–81
Avoid, Shift, Improve (ASI), 186

Bolivarian Revolution, 194
balloons, 1
bicycles, 122, 135, 152, 160, 188–89, 198, 207, 211–12, 214,
bipedalism, 1
biofuels, 87, 101–2, 109–10, 176
boats, 2, 97, 99, 109, 143, 182, 206
bohemians, 213
bourgeoisie, 39, 89
bus rapid transit (BRT), 73, 81, 178

Cadillac, 16, 33, 47, 49, 58, 158
cancer, 116, 124,164
car sharing, 174–76, 189
car-free cities, 206
Castro, Fidel, 158
catalytic convertors, 86, 90, 95, 100
caravan parks, 56, 61–62
cardiovascular disease, 114–15, 120
Centers for Disease Control, 93, 124
Chifley, Ben, 147
Children Against Road Rage, 133
Chomsky, Noam, 171
class: business, 49; capitalist, 27; working class, 31, 39–40, 47, 49, 55, 58, 67, 72–73, 120, 145, 155, 162, 166, 169
class conflict, 31

247

climate activists, 214
climate change, anthropogenic, 2, 4, 9, 43, 85, 87, 96–97, 101, 110–11, 164, 174, 183, 186, 190–92, 196, 198–99, 214, 216–17
Climate Council, 109
Clinton administration, 93
Coalition for Vehicle Choice, 94
Communist Party, 20, 34
Congestion, 3–4, 6, 8–10, 19, 52, 65–67, 72, 75–78, 80, 82, 89–91, 95, 110, 130–32, 144, 152, 157, 163, 173, 175, 182, 198, 200
Carbon footprint, 109, 205
Corporate Average Fuel Economy (CAFÉ), 197
Corvette, 55, 62
Critical Mass, 182, 185
cruise ships, 109
Cuban Revolution, 158, 194
custom automobile groups, 46
cycle rage, 211

disruptive technologies, 176
drones, 181

e-mobility, 178–79, 213
eco-driving, 168
eco-psychology, 207
eco-socialism, 192–95, 198–99, 216
ecological footprint, 126, 208
Electric Vehicle Association of America, 15
electric vehicles, 15, 177–78
Embarcadero Freeway, 173
Environmental Protection Agency (EPA), 97–99, 138
European Union, 43, 103
extraction, 3, 85, 88, 97, 100

fast food, 62–63, 126
fatalities, motor vehicle, 4, 9, 110, 116–23, 131, 133–34, 164, 181, 200
fascist regimes, 3–-33
Federal Aid Highway Act, 28

Five Year Plan, 20, 175
flying car, 181
food miles, 215
Ford, Henry, 15, 26, 30, 38, 48, 67
Ford Australia, 147
Fordism, 6, 8, 15, 27, 33, 47
Fordismus, 30
Foucault, Michel, 5
Fossil fuels, 1, 108, 178, 183, 189, 196
Freiburg, 200, 206–8
Freund, Peter, 2, 45, 57, 73, 86, 117, 126, 133, 141, 170, 187, 201, 204

gender, 8, 46, 48, 54, 58, 62
gentrification, 205
German Democratic Republic, 139–41
Global Climate Coalition, 88, 109–10
global democracy, 193
green state, 98
Good Roads movement, 27, 65
Google, 179–80, 213
Gramsci, Antonio, 6
green cities, 170, 205–6
green jobs, 195–99
Griffith, Walter Burley, 142
Guevara, Che, 158

Hansen, James, 199
Harvey, David, 70
hegemony, 4, 40, 62, 78, 81, 166, 182–83, 190, 204
helicopters, 82
heli-pads, 82
hitchhiking, 125, 212–13
Hitler, Adolph, 10, 36, 136–37
Holden, GM, 143–44, 146
holidaying, 203. *See also* vacationing
hominines, 1
horses, 9, 113, 145, 214
Hummer, 18, 41, 182
Humvee, 35, 37, 41
hyperautomobility, 73

Indigenous Australians, 145
industrial capitalism, 8

infectious disease, 119, 126
Industrial Revolution, 7
International Association of Public Transport, 186
international combustion engine, 3, 6, 14–5, 91, 98, 176
International Monetary Fund, 4, 192

Jevons Paradox, 108–9, 141, 167, 198
junkyards, 103

Kennedy, John F., 69
Kenya Vehicle Manufacturers, 161
killer commodity, 4, 134

Lada, 20, 85, 140, 158
Lenin, V. I., 33
Liftshare, 175
Light Rail Transit (LRT), 200
Low, Nicholas, 199–200
low-riders, 58

McDonald's, 70
McNamara, Robert, 28
Marx, Karl, 2
McLuhan, Marshall, 32, 55, 168
Mercedes-Benz, 33, 139, 141
metabolic rift, 35
Miller, Daniel, 135
modernity, 5, 21, 53, 56, 61, 74, 81, 114, 155, 159
mobility justice, 216
Mothers Against Drunk Driving, 133
motorcycle clubs, 46
motorbikes, 34, 173
mopeds, 78, 83, 119
motorization, 171, 187
motorized violence, 134, 139
Mumford, Lewis, 15, 43, 69, 82, 123
Mussolini, Benito, 30

Nader, Ralph, 121, 185
Nano, 100
National City Lines, 68–69
new towns, 207

new urbanism, 11, 200, 205–8
nitrogen oxides, 9, 85, 89, 91–92, 134, 138
noise pollution, 4, 9, 110, 114, 124–26, 131
Nordhoff, Heinrich, 137

Obama administration, 29, 100, 191, 197
obesity, 4, 7, 9, 73, 114–15, 120, 126, 129, 134, 164, 188, 211
oil crisis, 17–18
ozone, 89–90, 101, 106, 124, 134

Pacific Electric, 68
Partnership for a New Generation of Vehicles, 93
peak oil, 87, 171
permanent revolution, 193
planned centralized economy, 193
plastic, 87, 103
respiratory disease, 9, 85, 89, 92–94, 116, 124, 138
Porsche, Ferdinand, 36, 136
post-car system, 176
post-revolutionary society, 6–8, 21, 33, 85
powered two-wheel vehicles (PTWs), 25–26
Presley, Elvis, 50, 61
privatization, 195
psychic well-being, 127, 130–31
public ownership, 179, 193, 195–96

railroads, 65, 69, 105, 108, 160, 215. *See also* trains
real utopia, 193, 206
rebound effect, 108, 167, 198. *See also* Jevons Paradox,
recreational vehicles (RVs), 59, 63
reggare, 57
reforms: non-reformist, 165–67, 183, 213; reformist, 165–68, 174–75, 190
renewable energy, 178, 193, 195, 197–98

Reverend Ike, 49
risk society, 115, 181
road construction, 4, 88, 97, 100, 131, 145, 148–49, 160
road rage, 9, 132–33
Roosevelt, Franklin D., 27–3, 36, 115
ruteros, 158

Schuller, Robert, 49
settlement patterns, 65–84
sexuality, 8, 46, 54–55, 59–62
Simpler Way, 11, 159, 204, 213, 215
Skoda, 22, 158
Sloan, Alfred P., 32–33, 47, 158
Sloanism, 47
sociability, 2, 50, 52, 59, 127–29, 131–34
Special Economic Zone, 79
Speer, Albert, 32
speed, 133, 136, 140–41, 165, 167–68, 180
speed limits, 122, 134
sports utility vehicle (SUV), 18–19, 71, 88, 93, 97, 99, 117–18, 138–39, 162, 182, 200
smog, 68, 88–90, 100, 108
Stalin, Josef, 20, 33–34, 193
state capitalism, 26
steel industry, 103
suburban sprawl, 3, 67, 70, 73, 134, 144, 148, 189, 208
suburbanization, 66–67, 69, 82, 142
sulphur oxides, 9, 85
Supreme Court, 69
sustainable cities, 205
sustainable development, 89, 199
sustainable trade, 195
sustainable mobility, 199

taxis, 143, 153, 158–59, 162, 166
Tesla, 177, 179, 213
Trabant, 139–41, 158
trains, 1–2, 9, 19, 30, 39, 66–67, 74, 83, 87, 100, 104–5, 109, 117, 125, 130, 148, 153–54, 166, 168, 176, 181, 188, 191, 198–204. *See also* railroads
Transamazonian Highway, 155
Trans-Siberian Railway, 21
teleconferencing, 205, 212
trams, trolleys, 69, 72, 80, 83, 100, 130, 143, 146, 148, 152–53, 166, 217
Trotsky, Leon, 193

Uber, 174
United Auto Workers (UAW), 29
urban planning, 142, 157, 186, 200, 206, 208
urbanization, 75, 154, 171
Urry, John, 5–7, 57, 87, 114–15, 120, 132, 170, 188–90
US military, 37

vacationing, 11, 203. *See also* holidaying
vintage automobile groups, 46, 50, 158
Virilio, Paul, 31, 53
Volkswagen, 33, 35, 88, 106, 136–39, 147, 155, 161, 174

walkability, 207
Wartburg, 140
West Side Highway, 173
Wilderness Act, 182
Wilson, Charles, 27–28
Wilson, Woodrow, 113–14
Wolfsburg, 136–37, 141
woman driver, 55
workers' democracy, 195, 212
World Bank, 72, 118, 160, 192
World Health Organization, 114, 118, 166–67
World Trade Organization, 192
Wright, Olin Erik, 192
Wright, Frank Lloyd, 67

About the Author

Hans A. Baer is Principal Honorary Research Fellow in the School of Social Political Sciences at the University of Melbourne. He earned his PhD in Anthropology at the University of Utah in 1976. Baer taught at several US colleges and universities on both a regular and a visiting basis. He was a Fulbright Lecturer at Humboldt University in East Berlin in 1988–1989. In 2004 Baer taught at the Australian National University. He has published 21 books and some 190 book chapters and articles on a diversity of research topics, including Mormonism, African-American religion, socio-political life in East Germany, critical health anthropology, medical pluralism in the US, UK, and Australia, the critical anthropology of climate change, and Australian climate politics. Baer's most recent books include *Global Warming and the Political Ecology of Health* (with Merrill Singer, Left Coast Press, 2009); *Global Capitalism and Climate Change* (AltaMira, 2012); *Climate Politics and the Climate Movement in Australia* (with Verity Burgmann, Melbourne University Press, 2012); *The Anthropology of Climate Change* (with Merrill Singer, Routledge, 2014; 2nd edition, 2018); *Democratic Eco-Socialism as a Real Utopia: Transitioning to an Alternative World System* (Berghahn Books, 2018); and *Urban Eco-Communities in Australia: Real Utopias or Market Niches?* (with Liam Cooper, Springer, 2018).

www.ingramcontent.com/pod-product-compliance
Lightning Source LLC
Chambersburg PA
CBHW020115010526
44115CB00008B/832